Continuous Improvement in Organizations

RIVER PUBLISHERS SERIES IN MULTI BUSINESS MODEL INNOVATION, TECHNOLOGIES AND SUSTAINABLE BUSINESS

The River Publishers Series in Multi Business Model Innovation, Technologies and Sustainable Business includes the theory and use of multi business model innovation, technologies and sustainability involving typologies, ontologies, innovation methods and tools for multi business models, and sustainable business and sustainable innovation. The series cover cross technology business modeling, cross functional business models, network based business modeling, Green Business Models, Social Business Models, Global Business Models, Multi Business Model Innovation, interdisciplinary business model innovation. Strategic Business Model Innovation, Business Model Innovation Leadership and Management, technologies and software for supporting multi business modeling, Multi business modeling and strategic multi business modeling in different physical, digital and virtual worlds and sensing business models. Furthermore the series includes sustainable business models, sustainable & social innovation, CSR & sustainability in businesses and social entrepreneurship.

Key topics of the book series include:

- Multi business models
- Network based business models
- Open and closed business models
- Multi Business Model eco systems
- Global Business Models
- Multi Business model Innovation Leadership and Management
- Multi Business Model innovation models, methods and tools
- Sensing Multi Business Models
- Sustainable business models
- Sustainability & CSR in businesses
- Sustainable & social innovation
- Social entrepreneurship and -intrapreneurship

For a list of other books in this series, visit www.riverpublishers.com

Continuous Improvement in Organizations

José Dinis Carvalho
University of Minho, Portugal

With the collaboration of Rui M. Sousa

Published 2023 by River Publishers
River Publishers
Alsbjergvej 10, 9260 Gistrup, Denmark
www.riverpublishers.com

Distributed exclusively by Routledge
4 Park Square, Milton Park, Abingdon, Oxon OX14 4RN
605 Third Avenue, New York, NY 10017, USA

Continuous Improvement in Organizations / José Dinis Carvalho.

Routledge is an imprint of the Taylor & Francis Group, an informa business

ISBN 978-87-7022-798-8 (print)
ISBN 978-10-0084-749-9 (online)
ISBN 978-1-003-37411-4 (ebook master)

Contents

Forewords

No more excuses!

This is a rare book! It's one of those books that pops up once every decade, and that easily replaces a hand or two full of other books you have on your shelves right now.

For our privilege, the author organizes his vast and methodical investigation into the "secrets" that the giants of the world industry have developed and perfected over the last 70 years and that have allowed them to become, sustainably, the best of the best, and sharing to us, giving us just the juice, in simple and clear language, in just nine well-structured chapters full of practical examples.

However, the book in your hands, dear reader, is not for everyone. If you believe that continuous improvement and excellence in organizations are essentially matters of experts, of management tools, of tested solutions ready to be implemented, then this book is not for you. If, on the other hand, you believe that continuous improvement and excellence in organizations are matters of vision, purposes and alignment, values and principles, respect, and valuing people and teams, then this is the book you have been doing wait. And even if you belong to the first group, if you are curious and willing to learn, I guarantee you will not regret your careful reading.

Whatever your reason for seeking to deepen your knowledge of continuous improvement and excellence in organizations, this book gives you the rationale, the historical background, the comparative perspective, the answers and clues you need, and the ride. it still has access to a careful and very relevant supporting bibliography. It is also very gratifying to see that University/Business cooperation really works, when there is the entrepreneurship of Professor Dinis Carvalho and the opening of companies, as can be seen in the numerous examples shared, in the first person, throughout the work.

The author mentions as great motivation for writing this book "the clear awareness that all organizations/companies have an infinite potential that is being wasted every day and postponed with losses for everyone. Companies are losing everyday opportunities to become more competitive, grow, better serve their employees, better serve their customers and suppliers and better serve the families and communities where they are located and integrated. Every day our economy is losing opportunities to grow and we are all, in general, losing the opportunity to have a better standard and quality of life."

In addition to fully agreeing with this portrait of business reality, this seems to me enough motivation to act, and act now. Continuing to lose competitiveness and postponing opportunities is not only a bad business that does not augur well, it is, above all, stupid because a lot has already been invented a long time ago, and the path to be followed has already been covered by the best, as the author very well describes us in the pages of this work.

By the way, it is important to emphasize that if the reader is one of those people who believe that the principles, models, and examples described in this book only work in large companies because they are large, then it is worth clarifying that you have seen the film in reverse: large companies, have become large because these principles and models work, and because they practice them with conviction and consistency.

The question that really arises, dear reader, is whether you have the motivation, determination, and courage enough to move into active practice?

I particularly like the simple way in which Professor Dinis Carvalho summarizes what a leader must have in order to have any chance of success when he actually starts to practice continuous improvement: 1st "that the leader fully understands the direction the company is willing to go and what will be necessary to go in that direction.", and 2nd, "that the leader believes in people", I genuinely add.

The good news is that the path of continuous improvement and excellence, at the bottom of sustained success, is not a path reserved for the predestined. We can all achieve it: learning from the best, and for that it is enough to use this true guide for continuous improvement and excellence in organizations that you can finally have in your hands, as a base and inspiration; learning from mistakes as a team; and fully trusting people and their talents. Success, according to Churchill, is just that: it is going from failure to failure without losing enthusiasm.

My experience of over 30 years in the industry, in multinationals in competitive sectors, and in several countries, allowed me to prove that processes, systems, and tools are important, but the purposes, values, leaders, sharing, passion and enthusiasm, people, and their dynamics are truly essential, and they are the only real long-term competitive advantage that companies have over their competitors. After all, if 100% of a company's customers, as well as 100% of suppliers, as well as 100% of employees, are people, if we, as managers, do not understand people first and foremost, then what the hell is that do we know about our business?

Saying that the business world is changing very quickly and very unpredictably is commonplace, but it is also an unavoidable and unforgiving reality that will destroy many businesses and significantly change many others.

Not everyone will survive, but as in life, those who have more chances are not the strongest or with the most technical or capital resources, but those who have clear purposes, for which they are willing to fight every day and every day. circumstances, as Viktor E. Frankl, psychotherapist and Auschwitz survivor, teaches us in his extraordinary book "Man in Search of Meaning".

Practicing continuous improvement and the pursuit of excellence in organizations, as a way of life, in the service of a clear and inspiring purpose, can be extremely powerful if consistently practiced every day.

Someone said that it is by doing that you learn what you need to know, in order to know how to do it. Therefore, dear reader, without your enthusiastic, systematic and persistent action in your company, together with your co-workers and business partners, the exhaustive and comprehensive research and all the experience of Professor Dinis Carvalho, very well structured and In this unique work, it will have to travel much longer than desirable, so that the much-needed transformation of the competitiveness of our companies and the improvement of our level and quality of life are a reality.

Of the huge menu of organizational transformation proposals, intervention tools, management philosophies that promise success, none is as powerful and secure in the long term, and gathers as much evidence in both the business and academic world as continuous improvement and the search for excellence, when understood and applied as the author recommends in this work.

However, it is very important that the reader is permanently alert and aware of the absolute need to have their feet firmly on the ground, in all situations: "don't assume that solutions that work in other companies will have to result in yours… don't simply limit yourself to copy … the reality of (your) company is very particular and there is no other like it", and don't forget that "the transformation process must be aligned with the speed of learning and no steps can be skipped".

Have the courage to follow the author's appeal and make continuous improvement and organizational excellence a true way of life, serving an inspiring and well-communicated purpose and direction, supported by small steps, taken every day, by all, supported by proven knowledge and experiences, learning together what works and what does not work in their concrete reality, never being satisfied, promoting the systematic search for waste, recognizing and stimulating the talents of their people, sharing experiences with others who they are on the same path, always with enthusiasm, passion and valuing the diversity of opinions, and you will see that the reward will be gratifying and lasting. Aristotle is said to have said "We become what we often practice. Perfection, therefore, is not an isolated act. It's a habit!"

No more excuses!

Jorge Ferreira
Administrator and CEO,
IKEA Industry Portugal, S.A.

The understanding of the subject that the book brings, called by the author as Continuous Improvement - CI, is not a topic of natural knowledge of those who engage in strategies and / or production or sales planning, but it is extremely important for them to achieve their goals. To paraphrase here the author when he exposes his vision and understanding of CI: - "... *it is an instrument that allows organizations to systematically, gradually and effectively reduce the distance between their state and state they want or dream of* being", it is possible to understand the potential force that such knowledge and implementation can result. The potential that each company has, for sure can be increased with the daily incorporation of this tool or methodology for some.

History shows us that despite the constant evolution of the means of dissemination or access to information, the obligation to reach such teachings in reliable sources and, why not, produced by competent people proven by their deeds and legacies, are still the most used and efficient. The exchange of experiences, like an exchange, having on one side the academic environment, holder of up-to-date knowledge, through its daily studies and scientific research, and on the other side, a vast laboratory of commercial diversities, where planning and production strategies are regular activities, shows to be opportune and propitious. The academic environment populated by researchers, advisors and professors is a credit to those who dare to share the knowledge acquired to those interested or in need.

The author has his education, including undergraduate, master's and PhD in the area of Production (manufacturing). Through teaching, supervision and guidance of people at different levels of training, he has increased his experience and knowledge on the subject. Through active practices and hands-on methodology it was possible to know the reality of the sector and productive companies. The knowledge and experiences lived in productive environments prove his expertise and support the elaboration of a document, materialized with this book to share and make available in a scientific way the CI methodology, either to beginners in the academic life or to managers at the highest levels in the productive environment.

The work brings the subject in an adaptive way to different scenarios and realities, presenting and suggesting the implementation according to the possibilities, without sticking to or inducing a standard model. I dare here to make a comparison by quoting a jocular example: "*how do we feel when we seek a healthy diet and we are recommended a list of food inputs that are impossible to find in accessible places? The suggestion was given, but what we need to obtain good results are far beyond what we can acquire or achieve…*"

I conclude by thanking the honor of being able to preface this work, for the knowledge that is being shared in an accessible way and with the certainty that it will contribute to the managers, planners and employees involved with the continuous improvement in organizations.

Prof. DSc. Rui F. M. Marçal
Brazilian Association of Production Engineering - ABEPRO

List of Figures

List of Tables

List of Abbreviations

CONWIP Constant work-in-process
DMAIC Define, measure, analyze, improve, and control
FCT Fundação para a Ciência e a Tecnologia
FIFO First in, first out
GM General motors
KPO Kaizen promotion office
KBI Key behaviour indicator
KPI Key performance indicator
LERC Lean Enterprise Research Centre
CI Continuous improvement
TIEM Toyota inspired excellence models
MIT Massachusetts Institute of Technology
NUMMI New United Motor Manufacturing, Inc.
OEE Overall equipment effectiveness
PBL Project Based Learning
PDCA Plan, do, check, act
SME Small Medium Enterprise
POLCA Paired-cell overlapping loops of cards with authorization
PPDT Productivity press development team
VAR Value added ratio
SMED Single minute exchange of die
TPM Total productive maintenance
TPS Toyota production system
VSD Value stream design
VSM Value stream mapping
WIP Work-in-process

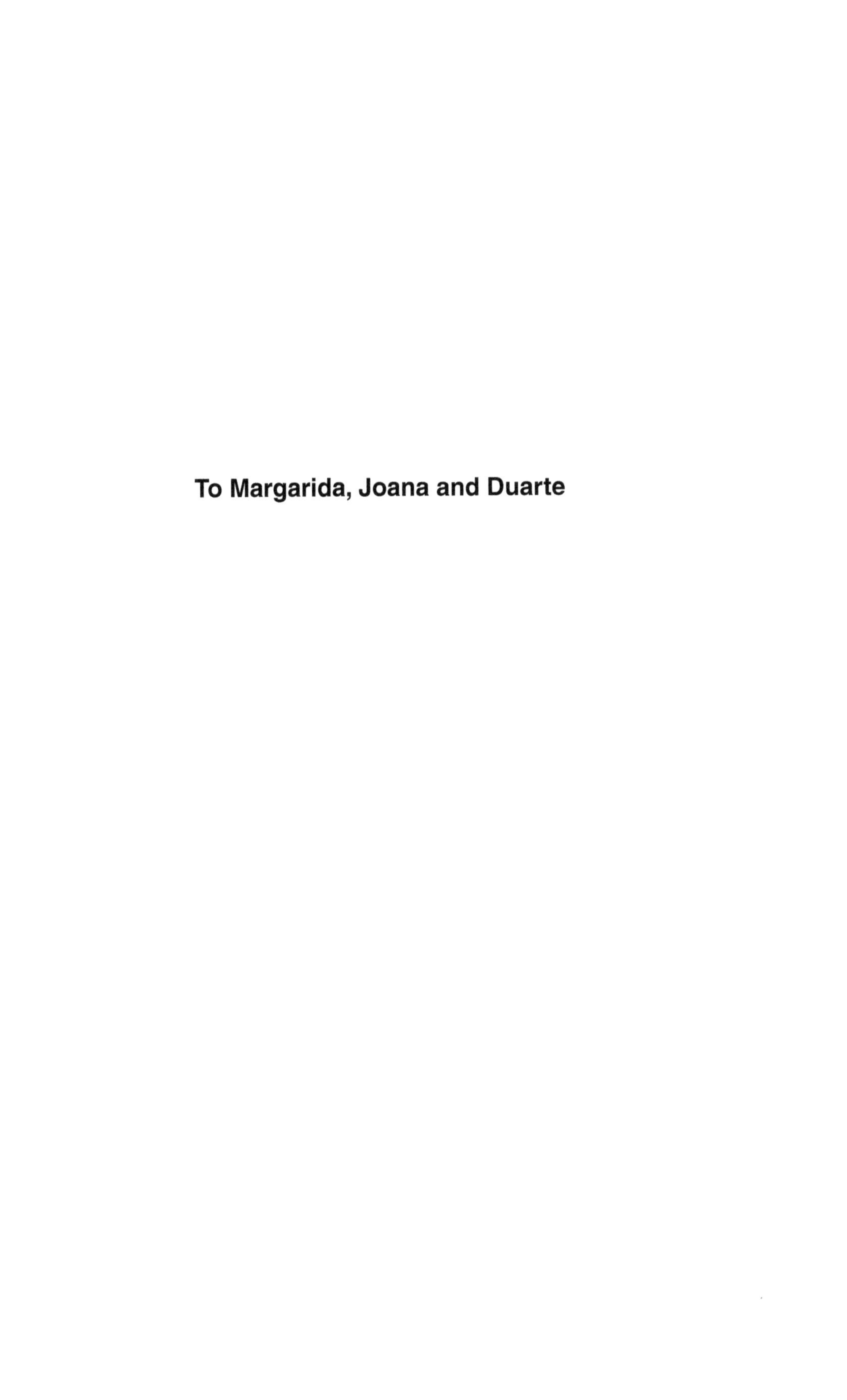

To Margarida, Joana and Duarte

Introduction

The main objective of this book is to share theoretical and practical knowledge that has been developed over years of study and practical experimentation in the areas of continuous improvement and excellence in organizations. A second purpose is to present the main organizations of the region where many interesting projects and experiences have been carried out and where some examples and practices are used as reference or as a way to clarify some concept or principle. Finally thanks will be given to people who have been important for this project to come to fruition and obviously a description of the way the book is organized.

Toyota's Approach

The mass production paradigm created mainly by Ford's assembly line, although very often criticized, had a very important role in improving people's living conditions in general. It is true that it also had clearly negative impacts but it was thanks to mass production that many people around the world had access for the first time to a wide variety of products. The assembly line made it possible to produce with much lower costs and with that many more people were able to buy them. Products that only the rich had access to, such as cars and some household appliances, became affordable for the middle classes in most countries. The impact of mass production was huge all over the world but in the 1950s a new production paradigm alternative to mass production started to emerge at Toyota. This production paradigm was called the "*Toyota Production System*" or TPS and was born out of the difficulty Toyota had to access markets big enough to take advantage of mass production. The challenge was to produce relatively low quantities of diverse products at mass production prices. The success of this production paradigm created by Toyota, although impressive, did not generate curiosity in the west until the 1973 oil crisis. This was due to the fact that after the end of the Second World War the western economies were growing in such a way that nobody felt the need to change the existing practices and tools. It was only when that crisis set in in the 1970s that many organizations, especially in the automotive sector, started trying to understand why Toyota and other Japanese organizations were getting through that crisis without any major problems.

The curiosity of the West about the production paradigm developed by Toyota began to have some widespread responses in 1977 in the first publication in English of TPS in an article in the International Journal of Production Research entitled "*Toyota Production System and Kanban System Materialization of just-in-time and respect-for-human system*" (Sugimori, Kusunoki, Cho, & Uchikawa, 1977). This article presents this new paradigm describing completely new principles, concepts and ways of thinking such as pull production, waste reduction, continuous improvement and respect for people.

More explicitly or more implicitly, several models have been created and promoted using the TPS production paradigm as a source, aiming at the search for excellence in organizations. In this book, we'll call them "Toyota Inspired Excellence Models" (TIEM). From these TIEM, the *Lean* philosophy (Womack & Jones, 1996), the *Shingo Model* (Plenert, 2017), and the *Toyota Way* (Liker & Franz, 2011) will serve as a reference in this book. The *Lean* philosophy for being probably the most popular designation, the *Shingo Model* for having been developed in collaboration with one of the creators of the TPS and the *Toyota Way* for being Toyota's own model of excellence. Besides these there are other known models such as *Quick Response Manufacturing* (Suri, 1998), *Agile Manufacturing* (Kidd, 1994), *World Class Manufacturing* (Schonberger, 2008), *Theory Of Constraints* (Goldratt & Cox, 1984), and many others.

Although the socio-technical nature of companies and organizations in general is a reality, Western organizations and academics have been more curious to implement and study the more technical side of these models, such as pull production, continuous flow production and *Just-In -Time*, than its more social and human side. As we are convinced that both sides have to coexist for excellence to be achieved, we decided to publish this book with a focus on continuous improvement and the role of people.

Motivation for This Book

Although the concept of Continuous Improvement (CI) is the key concept for the sustainability of companies and organizations in general, the truth is that it is not being properly recognized and understood by the overwhelming majority of our entrepreneurs and managers. It is very important that more and more people are aware of this and that steps are taken so that continuous improvement becomes popular in organizations, for the benefit of all. The continuous improvement is the tool that allows organizations to systematically, gradually and effectively reduce the distance that separates them from what they are now to what they want or what they dream to be. If you accept that to be true then it is important to understand as well as possible how to make the continuous improvement a reality in all organizations s and companies. The great motivation that gave the necessary strength for this book to be written was the clear realization that all organizations/companies have infinite potential that is every day wasted and delayed with losses for everyone. Organizations are losing every day opportunities to become more competitive, to grow, to better serve their employees, to better serve their clients and suppliers and to better serve the families and communities where they are located and integrated. Every day our economy is missing opportunities to grow and we are all generally missing the opportunity to have a better standard of living and quality of life.

In theory, if all our companies and organizations s become more competitive and better, in the light of what are the principles of continuous improvement and the principles of what we call Excellence in Organizations s, then our whole community will benefit. Employment would increase, wages would grow, working conditions would improve, job security and job satisfaction would improve, people would grow as individuals and, in short, the whole society would become better. This is more than enough motivation to at least try to share the knowledge we have about continuous improvement and Excellence in Organizations s with as many people as possible.

Over the last two decades, with the precious help of a group of colleagues from the Production and Systems Department at the School of Engineering in the University of Minho, we have been developing knowledge about Industrial Organization and Management through scientific research and practical experimentation in many organizations. Part of this developed knowledge has been mainly disseminated to the academic community through papers published in conferences and scientific journals. As we are professors, we have also been naturally disseminating an important part of that knowledge through classes and projects with students. Besides these channels, to which we have natural access in our profession, we think it is also worthwhile to disseminate the part of that knowledge that is more applicable in our companies and other organizations, by the community of professionals involved in Engineering, Organization and Industrial Management tasks.

Because we seek to develop knowledge through practical experimentation in a real context, we have been achieving an increasing interaction with organizations in the region and also with other types of organizations s such as hospitals and public organizations. This interaction has happened mainly in the areas of implementation of concepts, principles and practices popularly associated with the so-called *Lean* Philosophy, which in essence is a model inspired by the Toyota Production System. This interaction with companies and organizations s in general has been taking place in different approaches. One of these approaches, which often serves as a first contact with some organizations, is the guidance of students in their end-of-course projects, mainly in the form of Master's dissertations, which are carried out during a semester in companies with our supervision. PhD supervision is another similar format but with the difference that it has a stronger component in research than practical implementation in a real context.

There is another very interesting approach that has been adopted by our teaching team and that already has a large number of projects carried out in a real context in a Project Based Learning (PBL) program in which in our case we involve the 4th year students of the Integrated Masters in Industrial Engineering and Management. For over a decade now, our students have been developing semester-long team projects in companies with which we have partnership links. In these projects, each team of around 10 students carries out the analysis and diagnosis of one or more production units in a company, they identify problems and opportunities for improvement, design solutions and whenever possible implement them. These initiatives have been more or less constant in several of our partner companies in the region, from which we can highlight those where we have had projects more frequently, which are the following: APTIV (Braga), Leoni (Taipas), Bosch (Braga), Gewiss (Penafiel), Ikea (Paços de Ferreira), Preh (Trofa), Borgwarner (Lanheses), Balanças Marques (Braga); Continental ITA (Lousado), SNA Europe (Junqueira), Sonicarla (Mogege), Pinto Brasil (Guimarães). We would like to express our special thanks to these companies, but also to other companies with which we have had occasional partnerships.

Perhaps the most relevant approach that we have been assuming for interaction with the real context of companies is the interaction through research and development projects where our active participation is much more evident, with direct intervention in the field. Some of these projects have a strong research component, involving hired researchers and research fellows and are funded by bodies that promote science, such as the FCT (Foundation for Science and Technology). Other projects, which also include grant holders, are more oriented towards development and implementation in the field, are usually funded by the organizations themselves.

As a large part of these projects in their various formats have been bringing positive results for all stakeholders, we are having more and more partners. These projects play an important role, although in different ways, for each of the following stakeholders:

- For companies and other organizations s, because, in general terms, these projects challenge their managers and employees to alternative ways of thinking about the organization and management of their processes, routines and behaviors. This usually results in improvements in performance and in work conditions and satisfaction.
- For the students, because they get to know the real context and allow them to validate in a very effective way concepts, principles and tools with which they have had contact in classes.
- For the teachers, because they improve the effectiveness of their role in helping students acquire technical and professional skills while gaining and validating knowledge themselves. They also increase their networks of contacts in the business world.
- For the University in its responsibility to effectively serve the community around it, bringing to companies and other institutions new knowledge, new paradigms and new techniques so that they can become more competitive.

The nearly two decades of teaching, researching and field experiencing with Toyota-inspired models of excellence, such as the Lean philosophy, led us to feel that it might be useful to share with the community some of the knowledge we have been developing. In this area of knowledge, when you experiment in the field with real organizations and real people, is when the real learning happens. It is the famous "learning by doing". To be able to help someone to learn something, it is more than useful to have had field experience. This does not mean that field experience is enough to teach effectively, but it is very useful. There is no intention to show any kind of criticism to those who do it differently because we completely understand all teachers who do not feel stimulated by practical experience and prefer more theoretical research. We believe that we all have our part to play. We have always been attracted to the real context but believe me it is not an easy task. Going into the field is often difficult and sometimes even hostile. Owners, managers and employees in organizations have their demands and obligations and are not always willing to use their precious time to give attention to academics, as is our case. It is necessary to learn to communicate and to understand the side of those who have to pay salaries and fulfil their obligations to clients and to the country government. We have enormous respect for all those who have created their business and made it grow because we have gained the sensitivity to understand how difficult it is. And if the reader is someone who wants to implement continuous improvement in the company where they work or want to work, one piece of advice that we can leave, excuse the boldness, is to develop the ability to put yourself "in the shoes" of your interlocutor. This is called creating empathy. That interlocutor has to believe that you are genuinely interested in helping them and that you can actually do it.

With the successive field experiences, we became aware that the subject of continuous improvement is not being used by organizations in general and that with this we are all losing the opportunity to have a society with a better level and quality of life. We hope that at least some parts of this book may be useful to managers and industrial engineers that are involved in continuous improvement activities or may create some kind of motivation to start a continuous improvement movement in their organizations s.

Objectives of This Book

As knowledge about the subject has grown, either through practical experience in various organizations or through research and knowledge sharing with experienced practitioners and colleagues from other universities, it has become increasingly clear that the subject of continuous improvement is not well understood. In a large part of the community of academics, managers and decision-makers in our organizations, of the most varied types, neither the concept of continuous improvement is being well understood nor is its implementation being effective enough to take advantage of its potential. As continuous improvement is responsible for much of our future well-being, it is very important that managers and decision makers in organizations s know as well as possible this concept and that they develop competencies for its effective implementation. In the same line of reasoning our future as a developed and competitive society depends a lot on our ability to take advantage of this powerful concept and way of being that is the continuous improvement. This book intends in the first place to clarify the role of continuous improvement in the context of potential excellence in companies and organizations s in general and in their competitiveness and sustainability. Furthermore, it intends to clarify the role of continuous improvement and the Toyota-inspired excellence models in improving the well-being of employees, managers, clients, suppliers and the community at large. Secondly this book intends to show how the most successful organizations have organized and managed their continuous improvement processes and to show the reader how she or he can build continuous improvement in the small/medium sized company or organization where she or he works.

It would be very good to get the reader to accept with this book that there is no preconceived recipe to implement continuous improvement in an organization. There is not a pre-designed solution in the market nor a specific guide with all the steps for each organization. What you need is to understand that continuous improvement for your organization will have to be designed and conquered for your reality and context. But there is help you can turn to. There are some models available to serve as reference and therefore organizations can adapt one of them in the way that best fits their realities, history, market, or products, without being forced to follow faithfully an existing model. Nevertheless, many characteristics of the continuous improvement models that are implemented in organizations are shared by almost all of them. One example is teamwork and another is the monitoring of indicators in each team. This is just to give two examples. In this book we will present some models or meta-models of continuous improvement that will serve as reference for comparison with the models applied in the organizations selected and referred in this work. These models are Toyota Kata and Scrum which are published in books and some articles, while the information about the Kaizen Meta-Model was obtained through dissertations, web pages, some books and also from conversations with managers of the Kaizen Institute.

Local Benchmark Organizations

Some of the organizations we will use as reference in this book are mainly industrial companies, located in the northwest region of Portugal, and with which we have a very constant relationship, being our partners in various types of projects. These companies already have an interesting degree of maturity in terms of continuous improvement and excellence, some more than others obviously, and they continue to evolve in the constant pursuit of excellence. It is clear that in our region the organizations with formalized continuous improvement systems are a tiny minority, although in terms of maturity of their systems we can assure with great certainty that at least some of them are at the level of the best that is done in the world. Throughout the book we will refer here and there to examples taken from these organizations although in some cases we will preserve their anonymity by using fictitious names to designate them. These organizations are as follows:

- A manufacturing plant owned by a large multinational company located in our region, dedicated to the production of electronic equipment for the automotive industry, which will be referred to as *Company_C*. This manufacturing unit has more than 2,500 employees and with which the University of Minho has a very close relationship with large research projects, a joint doctoral program and many master's degree projects. This is perhaps the company that receives more students from our School of Engineering. This manufacturing unit is very mature in terms of continuous improvement, with great influence of the Kaizen meta-model, and is perhaps the company in the region with the most experience in this area.

- A factory in the furniture sector owed by a large multinational company that we will be referred as *Company_J*, employs around 1,300 people, very mature in terms of continuous improvement and has a very interesting culture focused on people. With this manufacturing unit we have maintained several research and development projects and some projects with student teams. Almost every year some of our students are also accepted to carry out their master's dissertations in their premises. This manufacturing unit has a very specific continuous improvement model, developed internally with great emphasis on the motivation and well-being of employees, managers and partners.

- A production plant owned by an American multinational company that we will refer to here as *Company_B* dedicated to the manufacturing of metal containers for the storage and transportation of gases under pressure. This company receives some of our students for master's dissertations and has also received teams of students for academic projects. This plant is another example of a company with its own continuous improvement system. The continuous improvement system in this plant started by being influenced by the *Toyota Kata* model, but evolved to a very particular model developed in the group.

- A production plant owned by an Italian group that is dedicated to the production of electrical devices and with around 600 employees. This company, here referred as *Company_H*, also receives our students to develop their master's dissertations and every year it receives at least one team of students to carry out projects in the "*Project Based Learning*" format. This plant has developed its own continuous improvement model over the last 20 years, which has given very good results in terms of performance improvements and its sustainability and stability.

- In order not to be limited to multinational companies, we will also include here *MoldartPovoa*, which is actually a SME with just over 100 employees and is located in Póvoa de Varzim, a city in the north of Portugal. MoldartPovoa is a company with an organizations al and management culture similar to the standard of Portuguese SMEs, but that slowly is implementing its own continuous improvement routines and pursuing organizations al excellence. The descriptions of some failed and some successful experiences in this company can serve as inspiration for all those who, working in SMEs, feel motivated to build their continuous improvement systems towards organizations al excellence.

- Also a national organization, but from a completely different sector, we include in this book the example of a public organization for solid waste treatment, located in Ermesinde (near Porto city) and called *Lipor*. This organization has been making a very positive evolution of its continuous improvement system with special relevance in the indirect and administrative areas. Lipor's continuous improvement system that has been evolving to adapt to the characteristics of public companies is very much based on the *Kaizen* meta-model.

- Finally we also decided to include another organization from a quite different sector which is the software development sector. The organization in question is here called *Company_P* with which we do not have joint projects, but for being from a different area we wanted to mention here. *Company_P* has its continuous improvement system very much based on the *Scrum* model of agile management. Actually, this organization has been evolving from Scrum to an approach called *DevOps* that besides Scrum includes some more concepts and principles that are part of the main excellence models inspired by Toyota such as the *Shingo model*, the *Lean philosophy* and the *Toyota Way*. The idea of *DevOps* is to include the adoption of principles such as "Make Value Flow and Pull", "Pursue Perfection", "Respect your network of partners and suppliers by challenging and helping them to improve." or "Become an organizations that learns through relentless reflection (*Hansei*) and continuous improvement (*Kaizen*)", just to mention some examples. The perception we have is that software companies that for some times have moved a bit away from the origins of *Scrum* (which was Toyota) are starting to approach that origin again with the adoption of the *DevOps* approach. The inclusion of this organization seemed to us very relevant because software development companies are in a highly competitive market and the way they organize and manage themselves has been quite different from companies in industrial areas. The reason for this difference is the intangible nature of their products and the constant need for creativity and innovation.

Acknowledgements

Many are the people and organizations s that contributed to make this book possible. In addition to the companies we have already mentioned in this chapter, I would like to express my special thanks to my longtime friend and colleague, Rui Sousa, from the University of Minho, for his invaluable help in proofreading most of the texts. I would also like to thank my colleague and friend José Barros Basto, professor at the Faculty of Engineering of the University of Porto, and my longtime colleague and friend, Venceslau Correia, professor at the Polytechnic Institute of Porto.

Structure of the Book

Chapter I will be entirely devoted to describing the origin of the thinking in which the concepts and principles of continuous improvement are embedded. The philosophy created by Toyota originates from the culture of the region of Japan where Toyota was born, although some credit must also be given to knowledge and ways of thinking that were imported to Japan from the West. Furthermore, this chapter also describes some key aspects of behavior and culture which shape the Toyota paradigm.

The concept of Excellence in organizations s is discussed in chapter II as well as the example of 3 models that are presented as having been inspired by Toyota. These 3 models are the *Lean philosophy*, the Shin*go Model* and the *Toyota Way*. The Lean philosophy was chosen because it is very popular and, in a way, responsible for the widespread promotion of the Toyota paradigm. The *Shingo model* and *Toyota Way* were chosen because they both represent in a good way the awareness of how the socio-technical nature of organizations s is vital to their success. In this chapter special reference is given to the fact that, in pursuing excellence, managers must be aware that apart from the visible (physical) side of organizations, which is a technical side, there is also an invisible side, which is the social science side, which cannot be ignored. It is important to realize the importance of each side and the ability to deal effectively with both is one of the keys to long term success.

Following what is presented in chapter II, chapter III goes deeper into the invisible side of organizations s while chapter IV is entirely devoted to the more technical, physical and visible side. Since there is a lot of content on this more technical side of the Toyota-inspired excellence models, this chapter IV is dedicated in particular to some aspects of flow management which do not appear easily in searches which might be made in books, articles or on websites.

The previous chapters have served to clarify the concepts related to the Toyota-inspired excellence models while chapter V is the first one specifically dedicated to continuous improvement. In fact, continuous improvement has to be framed in excellence models in order to make sense, so before going deeper into the subject it was very important that the whole paradigm of thought in which it is framed needed to be known first. This chapter aims to clarify what continuous improvement is about and the importance of its integration with the organization' strategy and its suggestion systems.

In the next chapter, the chapter VI, a classification model is proposed to evaluate the maturity level of the continuous improvement systems adopted in organizations. The number of proposed maturity levels is 4 and are evaluated in two dimensions, the visible dimension or performance and the invisible dimension or behavior.

Chapter VII is dedicated to describing how continuous improvement is put into practice according to classical implementation models. The models considered in this chapter are the *Toyota Kata*, the *Kaizen Model* and the *Scrum Model*. Although the *Scrum Model* is not often considered in industrial companies, the reason why this model is included in this chapter is because it comes from the same source and it can be actually applied successfully in many services of many companies.

The last chapter, the chapter VIII, presents a set of practical tips and suggestions that the reader should follow to implement continuous improvement in his company or organization. The chapter also describes some examples of practical implementations in some existing companies.

Finally, the reader can find in the annex short description of classical tool that are referred in the book associated to TPS and Lean. The descriptions tend to reflect the point of view of the author and also include suggestions of further reading.

References

Goldratt, E., & Cox, J. (1984). *The Goal: A Process of Ongoing Improvement*. Great Barrington: North River Press.

Kidd, P. (1994). *Agile Manufacturing: Forging New Frontiers*. Addison-Wesley.

Liker, J., & Franz, J. (2011). The Toyota Way to Continuous Improvement: Linking Strategy and Operational Excellence to Achieve Superior Performance. McGraw-Hill Publishing.

Plenert, G. J. (2017). *Discover excellence: an overview of the Shingo model and its guiding principles*. New York: CRC Press.

Schonberger, R. (2008). *World class manufacturing: the lessons of simplicity applied*. Free Press.

Sugimori, Y., Kusunoki, K., Cho, F., & Uchikawa, S. (1977). Toyota production system and Kanban system Materialization of just-in-time and respect-for-human system. *International Journal of Production Research*, *15*(6), 553–564. https://doi.org/10.1080/00207547708943149

Suri, R. (1998). *Quick Response Manufacturing: A Companywide Approach to Reducing Lead Times*. Productivity Press. Retrieved from https://www.routledge.com/Quick-Response-Manufacturing-A-Companywide-Approach-to-Reducing-Lead-Times/Suri/p/book/9781563272011

Womack, J., & Jones, D. (1996). *Lean thinking: Banish Waste and Create Wealth in Your Corporation*. New York: Fee Press.

1

The Origins

The search for better ways of doing the same thing has always existed. It's a bit of human history. If we are where we are today as a species, it is because we have improved over time. This chapter aims to place continuous improvement in the context of the historical evolution of industrial engineering and management and its role in the excellence of organizations. Some historical evidence will also be presented that may have helped precipitate the development of continuous improvement practices in organizations. As the evidence points to Toyota as the major driver of continuous improvement, the reader will be led to the origins of continuous improvement and its framework in the search for excellence in organizations.

1.1 Historical Framework of Continuous Improvement

The approach to production, or production philosophy, which began to be developed at Toyota Motor Corporation during the 1950s, represents, from the perspective of many industrial engineers and managers, the last major paradigm shift in the way of "thinking the production". The previous major milestone had been the beginning of the so-called mass production, materialized with the creation of production and assembly lines. This paradigm was developed in the beginning of the 20th century at the Ford Motor Company (interestingly another automobile construction company) in the United States of America, and it revolutionized industry and society at the time. Mass production brought a huge reduction in production costs, allowing the population in general have access to products that, until then, did not have. The production line was a fantastic change in the way of producing goods because it achieved impressive improvements in terms of productivity and, consequently, significant improvements in the living conditions of the population. Naturally, there were also aspects of this way of working that turned out to be less positive, especially with regard to the way operators were perceived by the system. In fact, on assembly lines, the human operator was regarded as a purely mechanical resource, i.e., capable of repeating small operations over long periods of time. This repeatability was seen - and still many people think so - as a way to achieve high efficiency gains, since it is admitted that the repeated execution of an operation by an operator leads to mastery in this execution. However, reality and accumulated experience revealed the existence of serious problems in this approach, namely: (i) higher occurrence of errors (leading to quality losses), (ii) higher risk of accidents, (iii) higher risk of developing musculoskeletal injuries and (iv) greater propensity for lack of motivation at work (due to the absence of challenges in repetitive work).

After this first major paradigm shift developed and implemented by Ford, with enormous success and acceptance by a part of the industry around the world, it is also an automobile company, this time Japanese, which develops the last important paradigm shift in terms of approach to production: the so-called Toyota Production System (TPS). However, although the first change (mass production) was, in general, well accepted by a large part of the industrial (and even academic) environment, the same is not the case with the second one, the TPS approach. In my opinion, since the end of World War II – when TPS began to be developed – until today, there has been no other disruptive change that in fact deserves to be presented as such, i.e., at the same level of the line assembly or the approach developed by Toyota.

Although it has taken advantage of the learning from all existing mass production knowledge and experience, the development of TPS has taken the way of "looking at production" to a quite different level. However, this new approach is often misunderstood as some of its principles and concepts are not easy to interpret and some of them even seem to contradict logic and intuition.

Until the early 1970s very little was known in the West about Toyota's practices and in fact very few people were interested in them. The reason was simple: From 1950 to 1970 western organizations and economies were growing steadily at high pace, economies were living the golden age, the GDP was growing in average 5% every year. It was the first oil crisis, which began in 1973, that changed all this. The price of oil increased 400% in a few months, causing a huge recession, both in Europe and in the United States, and consequently destabilizing the economy all over the world. It was at this time that the success of Toyota and other Japanese organizations aroused the interest of the West, whose organizations were facing enormous difficulties that even resulted in a dramatic number of bankruptcies. In fact, the West wanted to understand how Japanese industry could succeed at a time when others were languishing.

For a long time, many Western economic and political commentators tried to justify Toyota's success with Japan's cheap labor, with the idea that the Japanese worked immense hours, with the Japanese culture of a certain subservience at work, etc. This type of argument, used repeatedly in the 1970s and 1980s, was "comfortable" and convenient for the West, as it implicitly assumed that nothing wrong existed in the Western way of managing production. However, that "comfort" gradually dissipated in the face of a reality that became unavoidable: Toyota's approach was indeed innovative and effective. Since then, progressively, car manufacturers, both European and American, have been learning from Toyota, and those that did not ended up lagging behind.

Although Toyota had been using its new paradigm in its factories since the early 1950s, the truth is that the first publications about TPS in a scientific journal only appeared much later in 1977. With regard to books, both Taiichi Ohno and Shigeo Shingo published books in English about this new paradigm, its concepts and main techniques and practices, only in 1988 and 1985, respectively. Taiichi Ohno is known as the father of the TPS and Shigeo Shingo as being also involved in contributing to its development. Although Shingo's work (translated by Andrew Dillon), entitled "*A Revolution in Manufacturing: The SMED System*" (Shingo, 1985)", was published earlier, it does not present TPS in general, but rather a concrete methodology, designated by SMED (*Single Minute Exchange of Die*)[1], specially designed to reduce equipment setup times. Thus, the first English-language book on TPS was published by Taiichi Ohno, entitled "*Toyota Production System: Beyond Large-Scale Production*" (Ohno, 1988).

However, it was in 1977, that is, even before the publication of the books by Taiichi Ohno and Shigeo Shingo, that the first article in English about TPS appeared published in a scientific journal, the International Journal of Production Research (Sugimori, Kusunoki, Cho, & Uchikawa, 1977). This article presented several facets of TPS, including technical aspects such as calculating the number of kanban (way of controlling flow of products through the deferent processes)[2]. However, what aroused the

[1] SMED technique for the reduction of changeover time that will be briefly presented in the annex.

[2] Kanban is a technique to control the levels of inventory between processes and is briefly presented in the annex.

most interest in this publication was one of the two fundamental concepts of TPS. The authors state that TPS is based on two fundamental concepts, the first being "reducing the cost by eliminating waste" and the second "treating workers as human beings and with consideration". Although the first is "relatively" easy to understand, the same cannot be said of the second, even though, at first sight, it may seem trivial. In fact, none of the concepts is easy to fully understand and practice, but the second is far more complex. It is exactly this second concept that we will focus on more throughout this book.

1.2 Muda, Mura and Muri

For those less familiar with the concepts inherent to TPS, it is relevant to present the three great enemies of production, which are designated by the following terms in Japanese: *Muda* (waste), *Mura* (inconsistency) and *Muri* (overload), and are metaphorically represented in Figure 1.1. A renowned author of books on TPS, Hiroyuki Hirano, in his complete guide on *Just-In-Time* (Hirano, 2010) says that these "three enemies of production"[3] are known in Japanese factories as the "3 MU's" (they all start with the syllable "mu"). It is important that continuous improvement efforts constantly pursue not only the elimination of *Muda*, but also of *Mura* and *Muri*, since all of them have a negative impact on production in the short, medium and long term.

Muda corresponds to the concept of waste. Waste is understood as any activity or operation that does not add value to the product, but which, because it exists, consumes resources for the company. Seven types of waste were classified (Ohno, 1988): (i) movements, (ii) transport, (iii) waiting, (iv) overproduction, (v) waste in the processing itself, (vi) defects, and, (vii) stocks. More details about this waste can very easily be found in most *Lean* related material, whether it's books, youtube videos or websites.

The second enemy, *Mura*, which can be translated as inconsistency or variability, refers to all sorts of variations. Examples of *Mura* are: excessive load in one period and low load in another, processes overloaded and others with little load, people very busy and others not busy, sometimes the product is in compliance but sometimes not, as well as other types of variability. This variability, or inconsistency, is often considered to be the "nature of things",

[3] Although Hiroyuki Hirano has used the term "waste", both for the "3 MUs" and for one of them (*muda*), we have opted for the term "enemies of production" to avoid possible confusion with the term waste.

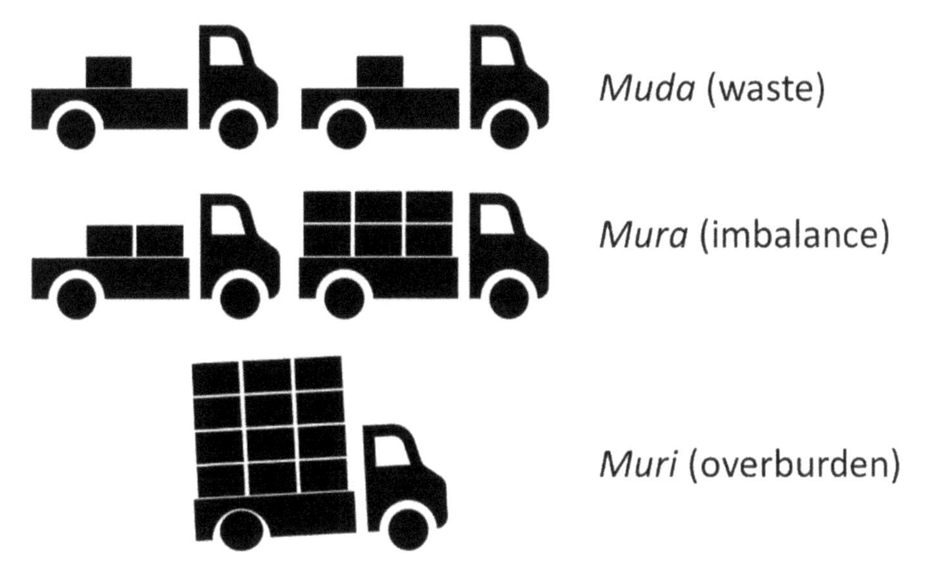

Figure 1.1 Metaphorical representation of Muda, Mura and Muri.

but in fact much can be done to reduce it. An example that is currently considered an inevitability is the case of variations in demand. In fact, there are variations in demand that can be caused by internal decisions of other departments or other companies in the same group. They are often also the result of ineffective communication with customers.

Regarding the third enemy of production, *Muri*, in Japanese it means something like physical stress or overload on people and equipment. The greatest focus in this book will be given to Muri on people since people are the most important resource of organizations and respect for people is one of the most important concepts for continuous improvement and for the success and sustainability of organizations. Therefore, any actions such as "bend over, bend over or stretch too far", "push hard", "lift heavy weights", "do repetitive tasks", and "take unnecessary walks" are considered *Muri*. There must therefore be a constant effort so that all these types of tasks are eliminated. These concerns were already present in the Toyota Production System, as shown in the first scientific article published in English in 1977 (Sugimori et al., 1977) on the focus that was given to people and the needs to protect and empower them. This focus on human factors has been maintained and affirmed at Toyota, as can be easily seen on Toyota's internet pages (Toyota_Europe, 2019) where "Respect for People" appears as one of its two main pillars, being the Continuous Improvement of other.

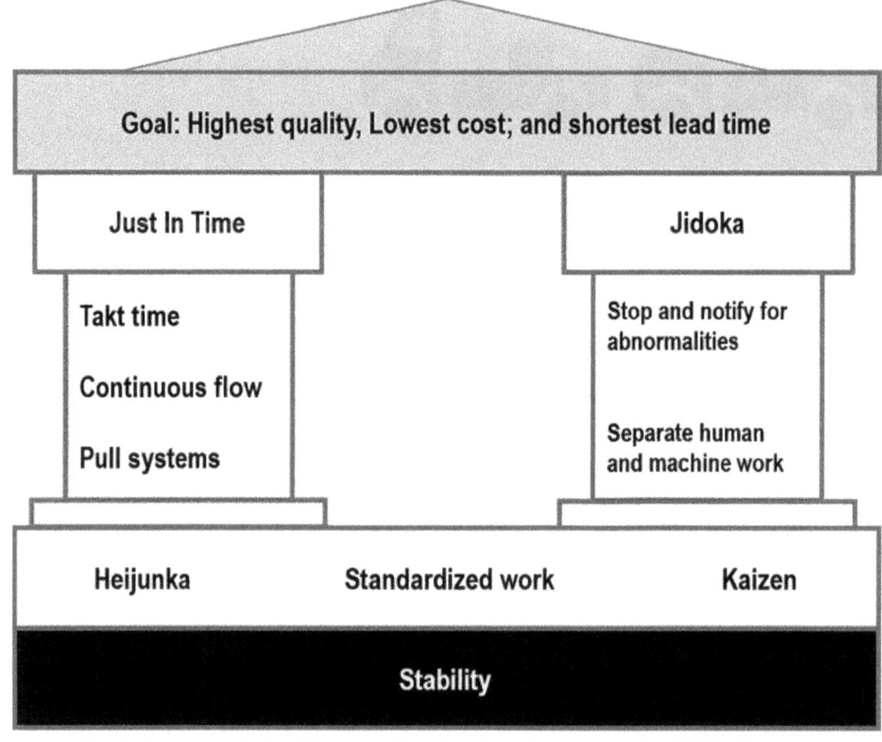

Figure 1.2 Original TPS house. Adapted from Narusawa & Shook (2009).

1.3 TPS House

Returning to the first book in English about TPS, Ohno indicates that TPS rests on two main pillars, called *Just-In-Time* and *Jidoka*[4]. Figure 1.2 represents the so-called "TPS house" where these two pillars clearly appear. This representation was first presented in the original version of the bilingual book (Japanese and English) written by Narusawa & Shook (Narusawa & Shook, 2009). Producing *Just-In-Time* is getting the right components and materials to the processes where they are needed, just when they are needed and only in the needed amount. It is important to understand that this concept would only

[4] *Jidoka* is a Japanese word applied by Toyota to the concept of equipping machines with systems that stop their operation when something wrong is happening. The word "autonomation" is also used synonymously.

materialize in its fullness in a perfect system, but that does not detract from its merit of showing us the way forward. We have to take constant steps in the way we organize and manage our production units in order to get as close as possible to the ideal *Just-In-Time*. There are many obstacles that hinder our progress towards pure *Just-In-Time*. Examples of these obstacles are:

- Equipment that, by its nature, produces several parts at the same time. Examples of this are ovens, stoves, stamping several identical pieces in each die, etc.
- Setups in a machine that force us to produce batches to spread the cost of machine downtime into several parts.
- Distances between machines that force us to create transfer batches (fill a box or container and then transport it to another location).

There are a number of other concepts and principles associated with the concept of *Just-In-Time* that it might be interesting to describe to make everything clearer. In these concepts we can include: *Takt* time, continuous production, pull production, one-piece-flow production, quick changeover in machine tools and total flow management. These concepts will be covered throughout the book.

Although a large set of versions of the "TPS house" can be found in Western literature, it may be pertinent to refer the proposal by Jeffrey Liker (J. Liker, 2004), illustrated in Figure 1.3. In this book, entitled "*Toyota Way: 14 Management Principles from the World's Greatest Manufacturer*", the author presents the "TPS house" in a more complex version that adds more information to the central part of the representation. However, this proposal by Jeffrey Liker is probably much more confusing (due to the greater amount of information) when compared to the simplicity of the original "TPS house". In terms of differences, the theme of continuous improvement is at the center of the house and not the base (or foundations), as it appeared in the original Toyota version (Figure 1.2). Another very relevant difference is the clear inclusion of human aspects, referred to in the figure as "people and teamwork", which did not appear in the original "TPS house". This inclusion by Jeffrey Liker of the aspects related to people in the "TPS house" seems appropriate, as it is consistent with the thinking of TPS and the Toyota Way Model, as we will see later.

The second pillar of the TPS house refers to the concept of *Autonomation*. This concept, originally known as *Jidoka* (Japanese word), was in fact introduced by Sakichi Toyoda (1867–1930), a great Japanese inventor and entrepreneur, initially dedicated to the manufacture of looms and later linked to

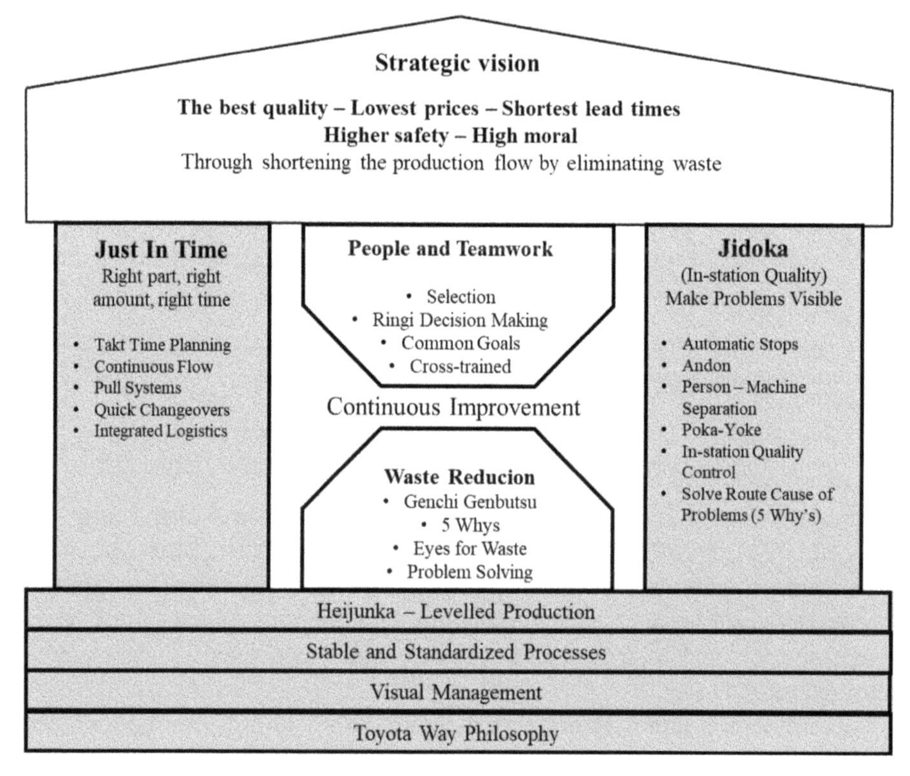

Figure 1.3 TPS house. Adapted from Liker (2004).

the founding of Toyota. *Jidoka* means that the machine has automatisms that make it stop automatically if any anomaly occurs. The idea is to introduce sensors and logic into the machines so that if an error occurs, the machine will automatically stop. This machine capability eliminates the need to employ a person who just monitors the equipment and intervenes in the machine if anything goes wrong. With autonomation in action, people could spend their time on tasks with greater added value where they can use more advanced skills than simply looking at an equipment.

As mentioned, of these two pillars of TPS, it was *Just-In-Time* that aroused the most interest and curiosity in Western managers in the late part of the 20th century. Publications on aspects related to *Just-In-Time*, e.g. production without stocks and kanban, began to appear in bookstores (especially after 1980) and in greater numbers than publications on any other topic related to TPS. Perhaps the first book on this subject, although without using the *Just-In-Time* designation in the title, was the one published in 1983 under the title "*Zero Inventories*" (Hall, 1983). The flow creation and stock reduction techniques aroused a lot of curiosity in the West because they were a visible part of TPS and clearly represented a disruptive paradigm, i.e., a way of thinking that contrasted sharply with the traditional thinking adopted by industrial organizations until then. Efforts to reduce stocks of materials in progress (*Work-In-Process* - WIP) was difficult to understand for those who were used to working with the concept of economic batch quantity and who believed that it was better to have stock to deal with the unforeseen.

The other aspects of TPS more related to the "human side" of production were always left to the background and only recently started to be the target of interest by western organizations. Even when the name Lean Production appeared (Krafcik, 1988), which became popular with the book "Th*e machine that changed the world*" (Womack, Jones, & Roos, 1990), the main focus was not on the aforementioned "human side" of TPS. It is also interesting to be aware that, even in the book that followed, entitled "*Lean Thinking*" (Womack & Jones, 1996), none of the five principles of Lean philosophy that the authors propose is dedicated especially to the "human side" or to "treating workers as human beings and with consideration", as referred earlier (Sugimori et al., 1977).

In this book, besides Lean, the Shingo Model and the Toyota Way will also be used as reference excellence models. Whenever the term Lean is used generically, it will be assumed to include the Shingo Model principles and the Toyota Way principles. In essence, the term Lean will include the whole movement that is shared by these three models. This generalization of the term Lean to the whole movement which was born with the Toyota Production System is assumed here because it is also assumed this way by a large number of professionals in the area.

1.4 The Origin of the Toyota Culture

Toyota's culture was influenced by the context that existed at the time and place where the organization was founded. The region where Toyota was established, Mikawa Prefecture, was a very rural region where the culture was very much linked to its ancestral heroes. To better understand how local culture influenced Toyota's culture, it is important to introduce three feudal heroes from the 16th century and 17th century in Mikawa and the neighbouring Owari region (Figure 1.4). In particular, the personality traits of Ieyasu Tokugawa marked the culture of the region.

Ieyasu Tokugawa, although the son of a nobleman, was taken hostage at the age of eight and remained so for twelve years. These adverse conditions sculpted his personality and helped him to develop some interesting

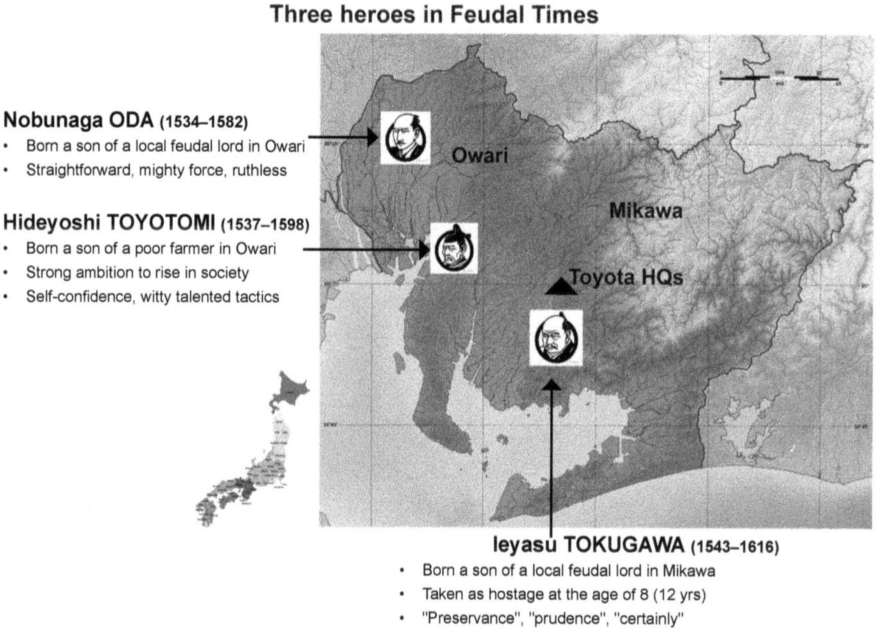

Figure 1.4 The three feudal times heroes from the Toyota region. Wikimedia Commons Files: Images in the Public Domain (https://commons.wikimedia.org/wiki/File:Oda-Nobunaga.jpg, https://commons.wikimedia.org/wiki/File:Toyotomi_Hideyoshi_(Kodaiji).jpg, https://en.wikipedia.org/wiki/File:Tokugawa_Ieyasu2_full.jpg e https://commons.wikimedia.org/wiki/File:Regions_and_Prefectures_of_Japan_2.svg [acedidos a 14/06/2021]); © Flappiefh (https://commons.wikimedia.org/wiki/File:Aichi_g%C3%A9olocalisation_relief.svg [acedido a 14/06/2021]). The use of this last file is regulated under the terms of the Creative Commons license – Attribution-ShareAlike 4.0 International.

characteristics, namely patience, learning from his own mistakes and diligence. There is a metaphor worth mentioning related to the personalities of these three heroes. Each one of them wanted a particular cuckoo to sing and, for that, the approaches they adopted were as follows:

- Nobunaga Oda said: "*If you don't sing little cuckoo I'll kill you*",
- Hideyoshi Toyotomi said: "*If you don't sing little cuckoo I'll make you sing*",
- Ieyasu Tokugawa said: "*If you don't sing, little cuckoo, I'll wait until you sing*".

The approach of Ieyasu Tokugawa, who is credited with some of the inspiration for Toyota's culture, is partially represented in this metaphor, which illustrates humility, patience, persistence and perseverance. To understand in more detail the personality of Ieyasu Tokugawa, we selected some of his famous quotes:

- *"Life is like a long journey with a heavy burden on your back. Take your time."*
- *"He who considers inconvenience to be natural will never be discontented."*
- *"When you want more than you already have, remember the days when you really had difficulties."*
- *"Patience is the basis of being safe forever. Consider anger as your enemy."*
- *"It is not good to know only the taste of achievement and not to know the taste of defeat."*
- *"Blame yourself and not others"*

Additionally, the region where Toyota was founded - Mikawa - is a mountainous area with unproductive soil and therefore a sense of communion, diligence, perseverance and loyalty to the local Noble was developed. Thus, Ieyasu Tokugawa's teachings and Mikawa's inherent context greatly influenced Toyota's culture. The result was a culture supported by values such as patience, humility, diligence/zeal, learning from one's mistakes, and stability. This was the kind of culture that was deeply ingrained in the people in the region where Toyota was established. These values were perfectly absorbed by Toyota resulting in a highly successful business culture to this day.

1.5 Leadership Practices from Toyota

There is a strong tendency for employees to say what their bosses like to hear, but at Toyota something different is practiced. The "bad news first" approach is encouraged by managers in order to more easily identify problems and find the root cause of those same problems. A "good boss" will find "WHAT to blame" instead of "WHOM to blame". So everyone is encouraged to bring problems to the surface.

In the common practices of our organizations, leaders make decisions based on information that comes to them in reports and presentations made by middle managers. At Toyota, leaders do not make decisions that way; instead, they go to the place (*gemba*) to see with their own eyes (*genchi genbutsu*) to feel the context and the problem there. Furthermore, it is expected that their judgment is also influenced by their "sixth sense", or intuition, to understand as correctly as possible what is actually going on in the place.

Setting seemingly impossible goals is another common practice at Toyota. 10% cost reduction is a type of goal that many organizations can set, but 50% cost reduction is already something that requires drastic measures. This helps to develop completely new ideas and new ways of seeing/approaching the same subject. This can be a powerful approach for any organization, as what seems impossible at first glance may actually be achievable.

> An example of a seemingly impossible goal was defined by us in a non-industrial organization in central Portugal. The problem that was presented to us was the following: a certain financial report, which had to be submitted quarterly to the state for reimbursement of relatively large sums of money, was taking between 30 and 40 days to be completed. This brought financial and liquidity costs that needed to be reduced. This long period (*Lead Time*) was a consequence of the time it took to gather information, which was collected in many different places and then compiled centrally. There were too many cases of people waiting several days for real relevant information (causing delays), confirmations/reconfirmations of information, too many late corrections, etc. The challenge that was proposed to us was to reduce the lead time to 20 days to complete the whole process, as this would already lead to a significant reduction in financial costs. However, we changed the challenge and set a goal of 3 days. At the beginning, there was some discomfort shown by many people because they thought we weren't "serious" or maybe we were too naive and we wouldn't be the right people for the job. After convincing them with some facts, we started to discuss with the team completely different ways of looking and analyzing the whole process. The truth is that the seemingly impossible goals were actually possible to be achieved.

Toyota also uses the practice of developing in people the ability to achieve goals. Goals are assigned to employees, but not great details or instructions are given for achieving them. Employees are encouraged to find the way to get the job done, and they know that whenever they need help, leaders will stop to listen and discuss with them.

Finally, another very important practice for success is to never being afraid of failure. This practice is in line with the ancient Chinese proverb "*Failure is the mother of success*". In the same line of thought is Thomas Edison's explanation when he tried and failed 10,000 ways to create an electric light bulb. He said "*I didn't fail, I just found 10,000 ways that don't work*". At Toyota the "*spirit of the challenge*" is as important as the result itself. Valuable lessons are

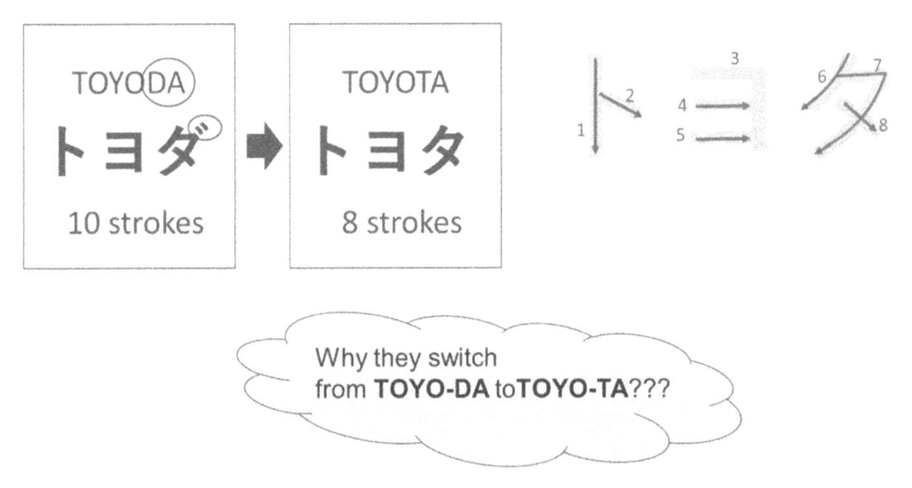

Figure 1.5 From Toyoda to Toyota.

learned from failure, which is why it is so important that people are constantly encouraged to take on challenges and not be afraid to fail.

1.6 Never be Satisfied

There is one ingredient that needs to be understood and embraced by all those who are dedicated to continuous improvement (the constant pursuit of perfection) if it is to be sustainable. That ingredient is "never being satisfied". Constantly pursuing perfection means never being satisfied with what has already been achieved and always seeking to do better. There is a very interesting Toyota story, on this subject, which was told to us by a former manager, Isao Yoshino, who was involved in setting up Toyota's first factory in the USA. We are referring to the NUMMI project (New United Motor Manufacturing, Inc.) which actually resulted from a partnership between Toyota and General Motors, and gave rise to the construction of a plant in Fremont, California in 1984. That car plant was in fact the first experience outside of Japan, of implementing the Toyota way of management and culture. Returning to Isao Yoshino, the story he told us did not convince us at first, but after a long and informal conversation we were eventually convinced. The story reveals one of the most fascinating aspects of Toyota's culture: The name Toyota comes from the word Toyoda, which already existed as a brand name (Toyoda Automatic Loom Works Company) and which is actually the name of the family that created that organization. In both Latin and Japanese spelling, the words Toyota and Toyoda only differ in a minor detail. However, it is in the Japanese writing that it is important to describe this difference: as can be seen in Figure 1.5, the writing of the two words differs only in two small strokes.

So, to write Toyoda you need 10 strokes and to write Toyota you only need 8. So why did they change the name from Toyoda to Toyota, since it's worth mentioning that even in this change a symbolic part of Toyota's culture is present? The first answer that occurred to us, and that may possibly come to those who are vaguely familiar with Toyota culture, is that movements are spared when writing the word (fewer strokes). While there is an understandable logic here, that is not the answer. The real reason, explained personally by Isao Yoshino, is that in many cultures, the number 10 symbolizes the top (the maximum) and so, according to this logic, 8 means being good, but still a little below the maximum, implying that improvement is needed to reach the top. This allegory aims to make all the people who work at Toyota understand that you are never good enough to the point that you don't need to improve. As Eiji Toyoda, one of the main people responsible for Toyota's success, used to say:

> *"Being satisfied with the current situation can be the first step towards corruption".*

The constant search for perfection (trying to be 10, assuming that you are always 8) is a kind of healthy "paranoia", based on the fear of letting ourselves relax with the successes of the past and thus increasing the risk of being overtaken by competitors. It is never allowing ourselves to accept that everything is fine, and the truth is that there is always something to improve, there is always some problem to be solved.

> On this same theme, Shigeo Shingo, frequently associated to the development of TPS, said the following:
>
> *"It is universal truth that those who are not dissatisfied will never make any progress. Yet even if one feels dissatisfaction, it must not be diverted into complaining; it must be actively linked to improvement."*

1.7 Keeping Continuous Improvement in Mind and in Daily Activity

Ono no Michikaze, or Ono no Tofu, lived between 894 and 967 and was an important Japanese poet and calligrapher of the Heian period (794-1185). Associated with Ono no Michikaze is an interesting legend - the legend of the frog and the willow branch - which still lingers in people's memories in various forms, notably on a playing card, a postage stamp and a small statue in the town of Kasugai, north of Nagoya.

Legend has it that on a rainy day, Michikaze, frustrated with the slowness with which he developed his calligraphic skills, decided to take a walk around the house where he lived. At one point, he saw a frog that tried several times to jump on a willow branch, but as the branch was too high, the frog failed every attempt. Michikaze began by concluding that the frog was wasting his time and energy, as it seemed he would never succeed. He continued to watch the efforts for a long time, but as much as the frog tried, he never succeeded. The frog eventually managed to reach the branch, because, with its successive failures, it learned some lessons and tricks to jump higher. This legend shows that if you don't give up and keep trying various approaches to succeed, you will learn over time. Michikaze learned an important lesson from the frog: not giving up and keeping trying would give him the opportunity to succeed.

In the English version of Wikipedia there is a slightly different interpretation, but this small difference profoundly alters the meaning of the legend: it is said that a breeze started to blow and that the branch bent and got closer to the frog that thus managed to reach it. Thus, it is assumed that the frog was successful by luck and not as a result of his effort, learning and dedication. Students in Kasugai City are taught at school that it was the frog's continued effort that made him jump into the branch; it wasn't by coincidence or luck. "Being diligent" is very different from "being lucky". Wikipedia can mislead readers by conveying the idea that the legend is just about waiting for a lucky break in order to win.

At Toyota great value is placed on the effort that is made by people to contribute small increments in a continuous and persistent way, as metaphorically portrayed in the legend of the frog. Reaching the willow branch represents reaching the goal, and, the frog's perseverance represents the continuous search for opportunities to improve. The concept and mindset of continuous improvement is prevalent throughout Toyota, not only in the people working in the factory, but also in all departments and at all hierarchical levels.

1.8 Suggestions System

An article published in 1999 in the scientific journal Total Quality Management, entitled "*Stuff the suggestions box*" (Lloyd, 1999), mentions an example of an ancient suggestion system which existed in Japan in the early 18th century, created by the eighth Shogun (high military rank) of the name Yoshimuni Tokugawa. This suggestion system consisted of a box placed at the entrance to his castle with the following sign: "*Make your idea known. Rewards will be given to ideas that are accepted*". Although many rewards were given to people who provided good ideas, proposals seen as critical of the Shogun himself often resulted in the beheading of the proposer. Modern suggestion systems began to appear in some organizations in the early 20th century, notably in American organizations such as Kodak and NCR, and Japanese organizations such as Kanebo. Later, but still before the Second World War, more precisely in the 1930s, organizations such as Hitachi, Yasukawa Electric and Origin Electric also introduced suggestion systems.

Toyota's suggestion system was introduced in 1951 under the name "Creative Ideas Suggestion System" and is described in some detail in the book "40 Years, 20 Million Ideas: Toyota's Suggestion System" (Yasuda, 1991). It is very interesting to note the importance Toyota already attached at the end of the first half of the last century to the creativity and innovation potential of employees. This was so clear that as early as 1953 they formally adopted the slogan "Good Thinking, Good Products". Not only the respect for creativity, but also the aspect of putting the customer first was well expressed in this motto.

The operators were encouraged and led to assume a very central role in solving problems that occurred in their work and also in developing suggestions for improvements. According to some internal Toyota documents we had access to, the attitude of the operators should be as follows:

> *"Once we have an idea of how best to run the operation, we first clarify the idea with our supervisors and then experiment and evaluate the results. We then fill in a creative suggestion and submit it for review."*

Each proposal is then verified by a supervisor or group leader, who assesses its usefulness, effectiveness and originality, using a standard scoring system to ensure fairness. Evaluations are carried out almost immediately and a cash prize is paid approximately two months after the suggestion is approved.

An alternative for forwarding suggestions uses the so-called quality circles, as can be seen on the right side of Figure 1.6.

A quality circle (or quality control circle) is a group of people (from the same area of work, or not) whose aim is to find solutions to solve quality problems or to improve quality. When a good solution is found, that same solution (or countermeasure) can be applied to all other places facing the same problem. When the members of these quality circles belong to different areas of production, the dissemination of good practices is made easier. Quality circles were very popular in the 1980s but are now rarely mentioned in organizations. It can however be accepted that today's project teams in continuous improvement which are not only dedicated to solving quality problems or to improving

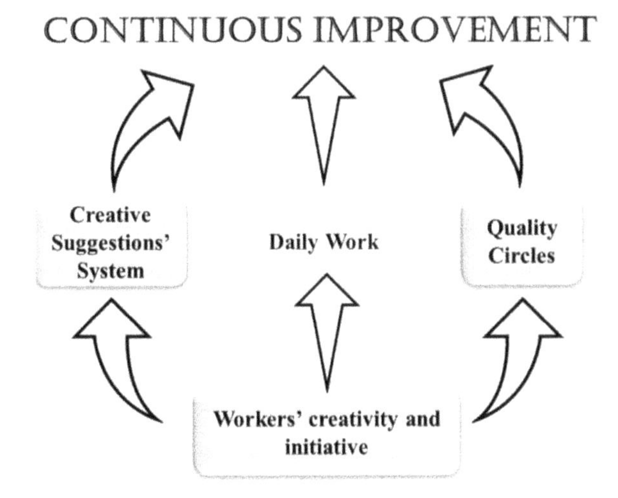

Figure 1.6 Ways in which Toyota employees can contribute to continuous improvement.

quality but also to other problems and opportunities for improvement in all areas, are a different version of quality circles. These project teams that promote disruptive improvements with process change are a very relevant entity in the structure of most continuous improvement systems.

The path "Daily Work" represented in Figure 1.6 concerns the direct contribution of the employees, for the continuous improvement, included in the daily work routines carried out according to the defined and updated norms, standards, guides and procedures.

In summary, there are three paths (not necessarily exclusive) of employee participation in continuous improvement. One occurs in the daily work itself (contemplated in the work rules). Another arises through the "individual suggestion" inserted in the Creative Suggestion System. The third consists of a "group suggestion" sent by the team members of the quality control circle.

Although it is assumed that Toyota's suggestion system was inspired by American suggestion systems (such as Ford's), the great effectiveness of Toyota's system lies in the emphasis the organization has always placed on its employees. In writings about suggestion systems, shared by our friend John Shook[5] in 2009, it is suggested that one of the faults of General Motors (GM) was clearly to neglect the human aspect in the suggestion systems.

[5] *John Shook is one of the most important personalities in the Lean movement. He is the author of several books on lean as the famous "Learning to See" dedicated to Value Stream Mapping (VSM). John Shook was involved in the first TPS application experience in America, the NUMMI project - an automobile factory that was born from a consortium between Toyota and GM.*

Table 1.1 Average number of workers per car produced.

	Toyota	USA	Sweden	Germany
Average number of employees per car produced	1.6	3.8	4.7	2.7

According to John Shook, GM decided that suggestions that did not save at least 20 USD would be immediately rejected because that was precisely the estimated cost of processing a suggestion. Clearly the system was focused on business interest and did not integrate the personal dimension. In general terms, conventional suggestion systems in America are geared to encourage BIG suggestions, to award BIG prizes, evaluated by BIG committees and with the expectation of BIG results. In contrast, Toyota encourages small improvements with small prizes. According to the author, most rewards at Toyota are around USD 10 maximum. Each employee submits on average one suggestion per week and 98% of the suggestions are implemented. Reflecting on the purpose of the Toyota system, the idea is not simply to improve processes; the idea is to improve employees' involvement in management and, as a consequence, to improve processes.

In the book "*A Study of the Toyota Production System*" published in 1989 (Shingo, 1989) the number of suggestions made at Toyota by its employees between 1976 and 1980 is reported: 10.6 suggestions per employee in 1976 and 18.7 in 1980 (with a 94% implementation rate that year). The same author adds that these suggestions have been responsible for the productivity advantage that Toyota plants have over their competitors. Still on this subject, a book published by Joakim Ahlstrom in 2014 (Ahlstrom, 2014), presents some interesting figures: 13 improvements implemented per person per year was a figure that very few organizations in Europe could surpass; and, 20 improvements implemented per person per year was the best that could be achieved in the world. The accuracy of these figures is difficult to guarantee, but they are indicators that can be used as a reference.

Table 1.1, relating to the late 1980s, compares a Toyota plant with other similar plants in America, Sweden and Germany with regard to the average number of employees required to produce one car.

Still on suggestions, the author of the book "*40 Years, 20 Million Ideas: Toyota's Suggestion System*" (Yasuda, 1991) describes the spirit of Toyota's suggestion system through the simple thought, "I want to make my job easier, even if just a little bit easier." Furthermore, he goes on to say, "If people remain alert to spot problems, in an environment designed to facilitate their identification, creative idea generation will not run out". The intent to frequently make small improvements is a very effective idea, but in light of the culture of most of our organizations, it seems a little counterintuitive. In our organizations we are always looking for a very good solution that will make us achieve significant performance gains and we don't give much importance to very small gains, sometimes even intangible, but which in the long term, because of their frequency, are extremely powerful. This is a key idea for the sustainability of continuous improvement.

In an article published in 2009, entitled as "*Human Resource development in Toyota culture*" (J. K. Liker & Hoseus, 2010), the authors state that the trust relationship created between the employee and the company or institution is extremely important and perhaps this is the reason why about 90,000 suggestions are submitted annually by Toyota's employees. It is also curious to note that, according to the same authors, Toyota intentionally does not emphasize monetary rewards. Generating ideas is supposed to be part of the job and as such you don't reward a person for just doing their job. The ideas generated are part of the trust relationship between the parties and that trust relationship is the currency of exchange between employer and employee. Employees can and should put their individual proposals in suggestion boxes installed at defined locations. They are encouraged to write them down or, if necessary, ask more experienced engineering staff to help write them down. They can also describe their suggestions verbally so that there is no justification for not submitting them.

Just to give an example of a organization with a large presence in Portugal, in a publication of the *Jornal de Negócios of 24 November 2014,* it is stated that Bosch in its factories in Germany saved 395 million Euros during a decade thanks to the suggestions of its employees. In 2013 the organization avoided spending 33 million Euros and rewarded employees with 7.7 million Euros for their suggestions (Prado, 2014).

Toyota handles the suggestion system in much the same way as it was used many years ago in the system that was originally implemented. There is a committee that evaluates all suggestions from employees. Whenever an employee has an idea for improving his work, he first talks about this idea with his direct boss (e.g. team leader or local supervisor) and then tries to implement (test) it to find out if it works. If the results/expectations are promising, the worker will put the idea into a specific pre-defined format, if necessary with the support of the boss, and formally send the idea to the committee. The committee evaluates the suggestion very quickly and decides whether or not it will be implemented and also what the reward should be. However, it is important to stress that the main goal of the system is NOT to give money to the employee who sends the suggestion. The main goal of giving a small amount

Figure 1.7 Cost-conscientiousness.

of money is to APPRECIATE the initiative of the employee and to encourage him/her to continue with this approach. Therefore the main objective of the system is not to reward the worker monetarily but to encourage his attitude.

1.9 Cost-consciousness

Awareness of the cost of things is deeply ingrained in the mind of Toyota's employees. They all understand that profit is the difference between the sales price and the cost of producing the product. They understand that the selling price is not the result of the sum of the cost of producing the product with an established profit. One example of this cost awareness is the practice of using the unprinted back of documents to make drafts and take notes (left side of Figure 1.7). Another example is found on the right side of the same Figure 1.7, and shows the solution adopted to use pencils that are too small to be used comfortably.

The employees of an organization only effectively care about the costs (small or not) of things in the organization when they are aligned and committed to the organization itself; the thing is that it is not at all easy to achieve this alignment / commitment.

References

Ahlstrom, J. (2014). *How to succeed with continuous improvement: a primer for becoming the best in the world.* McGraw-Hill Education.

Hall, R. W. (1983). *Zero inventories*. Homewood, Ill.: Dow Jones-Irwin.

Hirano, H. (2010). *JIT implementation manual: the complete guide to just-in-time manufacturing*. CRC Press.

Krafcik, J. F. (1988). Triumph of the lean production system. *Sloan Management Review*, *30*(1), 41–52. https://doi.org/10.1108/01443570911005992.
Liker, J. (2004). *Toyota Way: 14 Management Principles from the World's Greatest Manufacturer*. McGraw-Hill Education.
Liker, J. K., & Hoseus, M. (2010). Human Resource development in Toyota culture. *International Journal of Human Resources Development and Management*, *10*(1), 34–50. https://doi.org/10.1504/ijhrdm.2010.029445.
Lloyd, G. C. (1999). "Stuff the suggestions box." *Total Quality Management*, *10*(6), 869–875. https://doi.org/10.1080/0954412997280.
Narusawa, T., & Shook, J. (2009). *Kaizen express: fundamentals for your lean journey*. Lean Enterprise Institute.
Ohno, T. (1988). *Toyota production system: beyond large-scale production*. (C. Press, Ed.) (3ª Edição). New York: Productivity, Inc.
Prado, M. (2014). Bosch poupa 395 milhões de euros em 10 anos graças as sugestões de trabalhadores. *Jornal de Negócios*. Retrieved from https://www.jornaldenegocios.pt/empresas/detalhe/governo_espera_reforco_das_operacoes_da_bosch.
Shingo, S. (1985). *A Revolution in Manufacturing: The SMED System*. Oregon: Productivity Press.
Shingo, S. (1989). *A study of the Toyota production system from an industrial engineering viewpoint*. New York: CRC Press.
Sugimori, Y., Kusunoki, K., Cho, F., & Uchikawa, S. (1977). Toyota APAGAR. *International Journal of Production Research*, *15*(6), 553–564. https://doi.org/10.1080/00207547708943149.
Toyota_Europe. (2019). The Toyota Way: our values and way of working. Retrieved January 20, 2020, from https://www.toyota-europe.com/world-of-toyota/this-is-toyota/the-toyota-way.
Womack, J., & Jones, D. (1996). *Lean thinking. Lean Thinking*. Free Press. https://doi.org/9780743249270.
Womack, J., Jones, D., & Roos, D. (1990). *The machine that changed the world*. New York: Free Press.
Yasuda, Y. (1991). *40 Years, 20 Million Ideas: The Toyota Suggestion System*. Cambridge, MA, USA: Productivity Press.

2

Pursue Excellence in Organizations

Excellence has been desired, or pursued, by companies and other types of organizations through the application of concepts and principles, which in our opinion were created or strongly inspired by Toyota. Several designations have been attributed to "thinking models" originating in the Toyota Production System (TPS) but, in this book, emphasis is given to only three of them, specifically: (i) Lean Thinking or Lean Philosophy, (ii) Toyota Way and (iii) Shingo Model. These three are assumed to be reference models for the pursuit of Excellence in organizations because they seem to be the closest to the principles pursued by TPS. The so-called Lean Thinking, or Lean Philosophy, resulted from observing and studying Toyota factories around the world. The Toyota Way is the model assumed by Toyota itself as the evolution of TPS. The third reference model is the Shingo Model, which results from the vision of Shigeo Shingo himself, co-creator of TPS. To simplify the application of these reference models, special attention will be given to the principles presented by each of them, assuming that the principles largely represent the models themselves.

The body of knowledge in the area of Industrial Engineering and Management in which we are positioned is, by its nature, poorly structured, unregulated, and each scholar and professional uses the rules, terms and concepts that are shared in their environment. There is no entity, recognized by all, that dictates definitions, concepts and rules. This is certainly a weakness, but it probably results from the fact that it is a relatively new area of knowledge, the heterogeneous nature of sectors of activity and the huge number of protagonists worldwide. The result is an enormous difficulty in finding terminologies and definitions that are consensual and shared by all. Different players use the same term / designation for different concepts and different terms / designations for the same concept. Just as an example, and also to clarify the message we want to convey, let us look at the concept of cycle time. For some, cycle time refers to the time between the output of two successive parts from a machine or a line. This interpretation is shared by many Lean Thinking authors such as Mike Rother and John Shook in their famous book Value Stream Mapping (Rother & Shook, 1999). However, the same cycle time designation is used by other authors and contexts to define the time it takes for a product to go through a set of processes[1]. This ambiguity scenario gets even worse when, besides more quantifiable concepts (such as cycle time), we want to refer to concepts such as Lean, Kaizen, Continuous Improvement or Excellence in organizations. In fact, it is possible to find completely opposite interpretations, in which certain protagonists have considered these concepts to be the same while others consider them to be totally different. It is on these concepts or models that we will try to clarify a context and assume a point of view.

2.1 Models of Excellence in Organizations

The Toyota Production System (TPS) has inspired many models of excellence not only in production but also in the organization as a whole. Since the first journal article published in 1977 about TPS (Sugimori, Kusunoki, Cho, & Uchikawa, 1977) models have been created and evolving to the present day (see general overview in Figure 2.1). *Toyota Production System: Beyond Large-Scale Production*, was the first book in English about TPS (Ohno, 1988), published in 1988 by Taiichi Ohno, known as the father of TPS, although that version is just a translation of the first Japanese version published ten years earlier in 1978.

Models of excellence are understood here as being descriptions of how to proceed to achieve a competitive advantage in the market. In other words, they are descriptions of what to do, what principles to follow, and what tools to use to be more effective and efficient than competitors.

Probably the first kind of excellence model inspired by TPS, published in English after the TPS itself, was presented by Eliyahu Goldratt in his famous and bestseller book "The Goal" (Goldratt & Cox, 1984). One of the possible reasons that justify the success of this book is the fact that although it is a book with technical content it was written in a novel format. This

[1] In the book from Thomas Pyzdek (Pyzdek, 2003), in the page 707, the author states the following: "For example, one Six Sigma team working on improving purchase order (PO) cycle time (defined as the time from receiving a request for a PO to the time the requestor received the PO) ..."

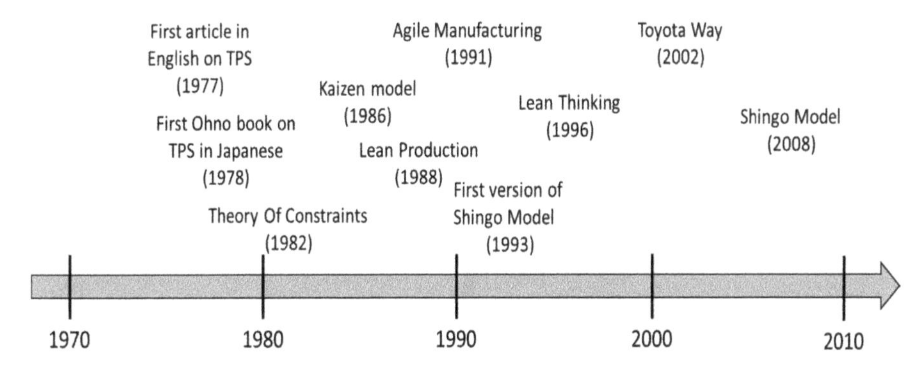

Figure 2.1 Main excellence models.

innovative way of presenting the model made it very attractive due to the ease of its reading and understanding. The model presented and coined as Theory of Constraints, became very popular as its Optimized Production Technology method was firstly published in 1982 (Fox, 1982) as well as the Drum-Buffer-Rope dispatching technique published a few years later (Goldratt, 1988).

Both Just-In-Time and Theory of Constraints models were very much focused on just one side of the socio-technical nature of organizations, the technical side, more precisely in the material flow control. "Just-In-Time" or just "JIT" together with "*Kanban*" has long been connoted in the West, in a relaxed way, as if it were the materialization of TPS or simply equivalent to TPS. Just-In-Time was referred in 1977 (Sugimori et al., 1977) and later referred by Taiichi Ohno (Ohno, 1988) as one of the two pillars of TPS. During these decades, most western organizations and universities were more interested in the physics concerning the flow control of materials than the human behavior and cultural side of TPS. JIT or "Just-In-Time" was accepted as a kind of operational excellence model pursued by most industrial engineering professionals and scholars. After the successful publication in 1990 of the book "The machine that changed the world" (Womack, Jones, & Roos, 1990) and later in 1996 with the publication of "Lean Thinking" book (Womack & Jones, 1996) the term JIT was gradually replaced by the term "Lean Production", "Lean Manufacturing", or simply "Lean". Although changing the term used, the Lean Thinking model was still very much focused on only the same technical side of the TPS as JIT.

The second excellence model inspired by TPS is most probably the one presented in a book by Masaaki Imai in 1986 (Imai, 1986). In that book, the author suggests that the economic success of Japan was the result of the Japanese management practices summarized in the so-called Kaizen umbrella presented in Figure 2.2. Under the umbrella, a list of concepts, principles, and

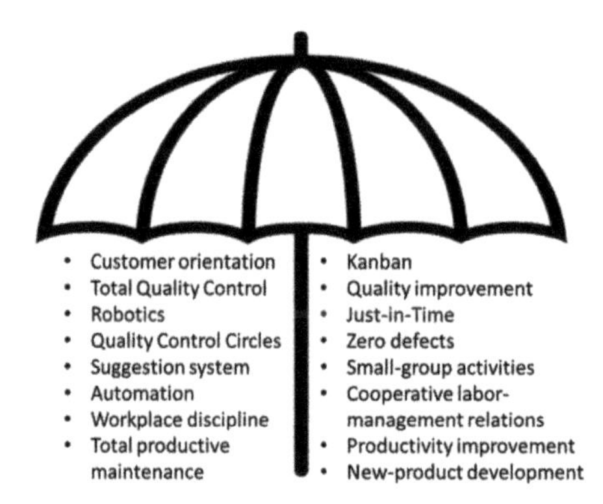

Figure 2.2 The Kaizen umbrella. Adapted from Imai (1986).

tools are presented as the Kaizen model guidelines or structure. From that list, it is possible to understand that the scope of the model covers the sociotechnical nature of organizations, from a more technical side to a more human side as expressed in the article referred earlier from 1977 about TPS (Sugimori et al., 1977). In that article, the authors argue that TPS is based on the following two main concepts: Reducing cost from the elimination of waste and treat the workers as human being and with consideration. In the items presented under the umbrella of Figure 2.2 the reader can see the technical aspects such as "robotics" and "*kanban*", as well as the human and behavior side as "Small-group activities" and "Cooperative labor-management relations".

Despite the existence of this very comprehensive model, during the 1980s and 1990s in the West, the terms that became popular were mainly "Just-In-Time" and "Kanban" as being the central part of TPS. Just-In-Time was referred in 1977 (Sugimori et al., 1977) and later referred by Taiichi Ohno (Ohno, 1988) as one of the two pillars of TPS. During these decades, most western organizations and universities were more interested in the physics concerning the flow control of materials than the human behavior and cultural side of TPS. JIT or "Just-In-Time" was accepted as a kind of operational excellence model pursued by most industrial engineering professionals and scholars.

This pulled approach to materials flow management, also often referred to as "Pull Production" is one of the principles that came from Toyota that has most attracted the attention and perplexity of the West in the last three decades of the 20th century. This concept goes against the traditional way of thinking that was never questioned in the way production was organized and managed. Take for example a proverb of popular wisdom which says "Don't put off till tomorrow what you can do today". Pull production is exactly the opposite, it is more like leaving for tomorrow what you only need to do tomorrow. This subject will be clarified later but understanding it remains a challenge for many professionals with responsibility for organizing and managing production and its flows in most of our organizations. If the reader manages to understand this concept well and accept it completely he or she will have already taken a big step towards mastering the subject of organization and operations management.

Despite the focus of Lean Thinking was on the technical side of organizations, such as value, value stream identification, and pull flow, the principle of pursuing perfection leaves some room for the social sciences' side. While the importance of teamwork, empowerment, motivation, and bottom-up initiatives are also briefly referred in that original book, the focus of Lean Thinking is towards value, flow and its continuous improvement. Lean Thinking was materialized as following 5 principles: (1) identification of value, (2) identification of the value stream, (3) promoting flow, (4) promoting flow pulled by demand, (5) pursue perfection (also known as continuous improvement).

Agile Manufacturing (AM) is another famous model of excellence proposed by a group of researchers at Iacocca Institute in 1991 (Nagel & Dove, 1991). This model comes to life shortly after the first scientific article presenting "Lean Production" (Krafcik, 1988) and the famous book "The machine that changed the world" from which Lean production became famous and just two years before the book "Lean Thinking" being published. Maybe inspired in TPS, the AM model clearly distances itself from the TPS questioning some of its concepts and never mentioning some of the classic TPS tools such as *5S*, *SMED*, *Heijunka*, *Kanban*, and *Poka-Yoke*. In this model, there is an important component of the inclusion of new technologies and in the integration of the following 3 pillars (Kidd, 1994): Organization, People, and Technology. The Organization pillar refers to the innovative management structures and organizations; The People pillar refers to the skill base of knowledge and empowered people, and the Technology pillar refers to the flexible and intelligent technology. The AM conceptual framework includes Competitive foundations, Core concepts and Generic features model, as described in Figure 2.3.

Although *Agile Manufacturing* did not manage to become popular in real implementation in organizations the term *Agile* ended up being used in a different context, in software development organizations. Many of the principles and

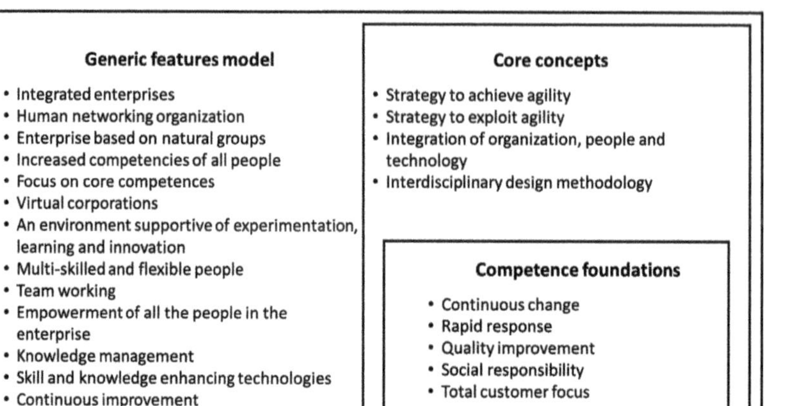

Figure 2.3 Conceptual framework of agile manufacturing. Adapted from Kidd (1994).

concepts created in TPS became through that word adopted in software application development, namely in the respect of teamwork, work visualization and workflow. The main tool associated to *Agile* is the *Scrum* methodology, not only popular in Software development but also in project and team management in different contexts. Currently, large software application development organizations are migrating from the *Agile* designation to the *DevOps* designation by reviving TPS principles and concepts such as WIP control (controlled flow), getting it right the first time and continuous improvement.

With the work of the MIT group led by James Womack, names that included the term *Lean* began to be used, namely *Lean Production*, *Lean Manufacturing*, *Lean Management* and *Lean Thinking*. The term *Lean* also appeared in other designations associated with different activity sectors, for example *Lean Hospitals*, *Lean Construction*, *Lean Office*, and *Lean Education*. The single term *Lean* has also been used to designate this line of thought created at Toyota. Finally, another term that is very often used as an equivalent to *Lean* is the Japanese term *Kaizen*.

In 2001 Toyota formalizes a set of principles and behaviors that were established and practiced in the company by naming them the *Toyota Way* (Liker & Franz, 2011). This statement of the Toyota model with the name *Toyota Way* represents the evolution of TPS to other areas of activity than just production. The way of thinking that worked so effectively in production proved equally effective in other parts of the organization, namely in administrative areas, product development and, indeed, in all other departments. The 14 principles of the Toyota Way have been grouped into two main dimensions that appear in the motto shown for example on one of the Toyota webpages

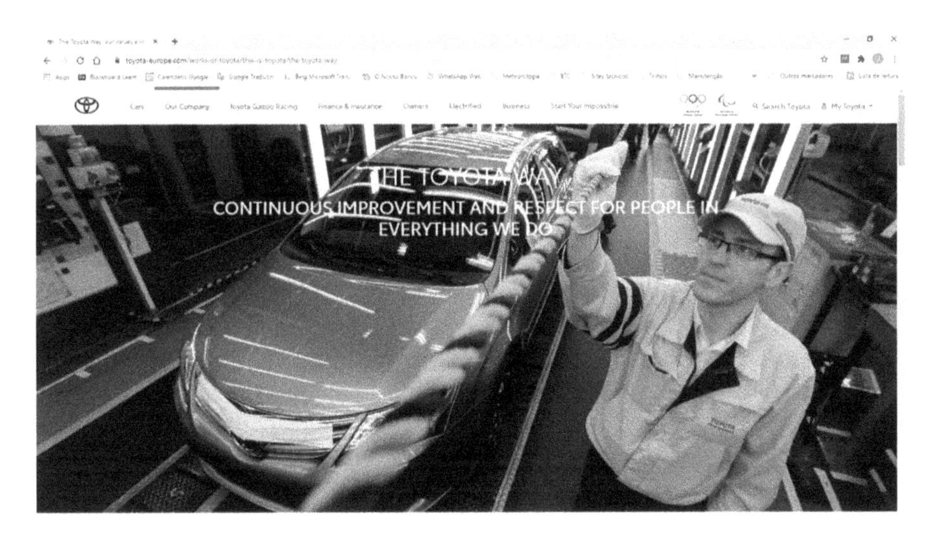

Figure 2.4 Motto at the Toyota Europe website.
Source: https://www.toyota-europe.com/world-of-toyota/the-toyota-way (accessed to 27/01/2021)

(Toyota_Europe, 2020): "Continuous Improvement and Respect for people in everything we do" (Figure 2.4). Jeff Liker, a famous Toyota scholar and author of several books on the subject, published a book in 2003 where he presents and describes these 14 principles of the *Toyota Way* (Liker, 2004).

On the other hand, another important interpretation of the way Toyota organizes and manages itself is presented by the *Shingo Institute* in what they have dubbed the Shingo Model (Shingo Institute, 2020). As those of you that are reading this book and working in this area of continuous improvement know, Shigeo Shingo also contributed to the development of the TPS (being Taiichi Ohno its main creator) and therefore this model was developed with the collaboration of a person who knew Toyota very well, Shigeo Shingo. The *Shingo Institute* frequently mentions in its publications that its model is a model for to pursue *Operational Excellence* and maybe that is the reason why many people associate the designation *Operational Excellence* with the Shingo Model.

> It is important to be aware that all organizations and organizations should constantly pursue excellence for the good of all around the world but in truth none can ever achieve it. Excellence in organizations is like a point in infinity that no matter how far we walk towards it, we will never get there.

In our interpretation, the designation *Excellence in organizations* can be the targeted (or be pursued) by the utilization of various models or philosophies, but this designation has no specific author or model. "*Excellence*" here is just a term of everyday language that in general all people understand. We are familiar to the word "excellent" as an adjective to refer to a dish served to us or to anything else. We can say that a certain person is an "excellent" speaker or an "excellent" professional. It is with this kind of interpretation that we want the term excellent to be understood here in this book. Next, we will clarify the meaning of Excellence in organizations that is assumed in this book.

There are three frequently published concepts that use the term *excellence* and are relevant in this context. These concepts are *Business Excellence*, *Operations Excellence* and *Service Excellence*. The relative position between Business Excellence and Operational Excellence is presented in an interesting article by Cindy Johnson of Texas Instruments (Johnson, 1997), in which Business Excellence is the highest level goal and Operational Excellence one of the supporting instruments. Regarding the roles of Operational Excellence and Service Excellence, Thürer, Tomašević, Stevenson, Fredendall, & Protzman (2018) suggest in their article that the former is related to efficiency and the latter to effectiveness. It actually makes some sense since "effectiveness" is getting the goal accomplished while "efficiency" is getting the goal accomplished but using less resources/effort/time.

There are several possible definitions for Operational Excellence but we will only mention here some that are aligned with our way of thinking, and they are the following:

- "Providing customers with reliable products or services at competitive prices and delivered with minimal difficulty and inconvenience" (Kamann, 2007);
- "The degree to which a firm is better than its competitors in responsiveness and generating productivity improvements" (Rai, Patnayakuni, & Seth, 2006);
- "Outstanding achievements in manufacturing or service processes and productivity improvements, quality improvements and customer service" (Chakravorty, Atwater, & Herbert, 2008);
- "Operational excellence in simple words is when improvements are made to gain a competitive advantage" (Sony, 2019).

These and other definitions almost always focus on the issue of improved performance in the production of goods or services. We could then assume that an organization has a higher degree of Excellence than its competitors if it has better global performance. This global performance includes performance such as productivity, product quality, speed in satisfying orders, service quality, low costs, motivated and satisfied employees, good relationships with suppliers, etc.

The models that have been selected to be presented here in this book, which when implemented and followed contribute to excellence in organizations, are the following: the *Lean philosophy* (*Lean Thinking*), the *Toyota Way* and the Shingo Model. The term "Excellence in Organizations" or just "Excellence" will be used in this book as being equivalent to "Operational Excellence". The reason is that the term "Operational Excellence" is for many, linked to the Shingo Model and on the other hand for many people only linked to industrial production, leaving out services and other processes in the indirect areas in organizations.

2.2 Principles of Lean Philosophy

Principles are guidelines that are established in order to condition behavior; serve to help determine whether or not a given behavior is appropriate in relation to them; and also help to select the tools/solutions that best align with the organization's philosophy. Principles are the ground rules of an organization and necessarily have consequences for decision-making and conditioning behavior. Each one of the excellence models presented in this book is shaped by its principles and for that reason understanding the principles is necessary to understand the model.

The five principles of the Lean Philosophy were presented in 1996, in the book entitled "Lean Thinking" (Womack & Jones, 1996), a little after the book "The machine that changed the world" (Womack et al., 1990) which made the term "Lean" popular worldwide and which is still a great reference in the world of Excellence in organizations. The five principles that guide the Lean philosophy are:

Principle 1. Specify Value,
Principle 2. Identify the Value Stream,
Principle 3. Flow,
Principle 4. Pull,
Principle 5. Perfection.

Each one of these principles will be described and discussed below.

Principle 1 - "Specify Value" - This principle consists in identifying, as far as possible, what the market interprets as value in the products the company offers.

The customer's point of view (regarding the product's characteristics) is more important than the point of view of those who design and produce the product. This idea is very much linked to the old saying "no one should be a judge in their own cause" from the Latin "*Nemo judex in causa sua*". The value of the product should not be assumed to be its total cost since many of the partial costs may result from the inclusion of operations that in reality the customer does not value or does not want. These costs are associated with the term waste (*Muda*).

The concept of value is an important one but may not be easy to define in a precise and consensual way. Value is also often described as everything the customer is willing to pay for. This description is not completely right because the truth is that the customer is frequently willing to pay for waste as well. Waste is paid for by customers otherwise organizations would not be able to survive. It is probably safer to say that the value of a product is in most cases defined by the market and therefore results from the balance between supply and demand.

In order to clarify the concept of value we will add another aspect that can increase the interpretation of the concept. We can say that an operation adds value to a product when it produces a physical or chemical transformation and that transformation is recognized by the customer as added value. In short, we can say that operations on a product can transform it physically or chemically, simply change its position by transport or handling, or not cause any change in the product. By definition, all operations that only change the position or simply do not result in any change in the product are understood as waste. This does not guarantee that operations that result in physical or chemical transformation represent value added. There may be physical and/or chemical transformations that are not recognized by the customer as added value and therefore cannot be considered as value adding operations but rather as waste. This type of waste is called "over-processing" and will be presented later in this chapter.

Much can be said about the concept of value as it is a complex concept but it goes a bit beyond the scope of this book. It is not normally the job of Industrial Engineering and Management professionals to address the issues of product value but rather the reduction of costs, lead times and defects associated with those same products. Our focus here is to clearly understand the difference between operations that add value and operations that do not add value (waste). If the operation that is being carried out on the product does not result in something that is recognized by the customer as "value" then we will be in the presence of waste.

The concept of waste (from the Japanese *Muda*), in contrast to the concept of value is a very important concept in *Lean philosophy* and is actually inherited from TPS. The concept of waste has long received special attention from some production thinkers. Henry Ford, for example, the father of mass production and the creator of the assembly line, already used this concept as a focus in his production methods. Taiichi Ohno strongly evolved the use of this concept during the development of TPS. The reduction of production cost was essentially based on waste reduction, assuming that anything beyond the absolute minimum of equipment, materials, component and labor for production would only exist to increase cost (Sugimori et al., 1977).

Waste can be understood as any activity in a production system that does not add value to products. It is more or less easy to understand that the identification and elimination of activities or operations that do not add value to products results in improvements on production performance. Waste present in every production unit, is always around us, but the truth is that it is not always easy to be identified by an inexperienced observer.

Let us assume the following situation. Just imagine yourself going into a production facility and looking around you observe the following:

- An employee carrying a visibly heavy roll of fabric.
- One employee actively looking for something in a drawer.
- Two employees heading towards the entrance of another production area.
- Two employees very attentively receiving some instructions from the foreman.
- An employee looking through the transparent protection of a machine in order to check if everything is running as planned in the operation.
- An employee carefully transporting a very full pallet on a forklift.
- An employee removing defective parts from a box.
- An employee stretching to reach a handle.
- The foreman delivering a paper manufacturing order to an operator.
- An employee waiting for another employee to finish unloading material next to the machine.

At first glance everyone is busy and very engaged with their work but in reality everything they are doing is wasteful. None of the people observed were adding value to the products and this happens more often than it might seem.

Let us look at some classes of waste:

- Transportation - The product is transported within the factory between the various sections. These transport operations uses organization resources, uses at least labor, some form of transport and quite naturally energy. All this expenditure does not result in any value recognize d by the customer. No customer will recognize value (and pay more for) a product that has moved more than another product within the plant.
- Storage - Storing goods also uses organization resources including labor, space and energy. Customers do not recognize value in products because they have been stored too long (there are some exceptions, such as whisky and old brandy).
- Rework - Performing an operation to repair or improve one product that has been carried out incorrectly, uses organization resources, represents a cost but does not add value to the item.

The reader may argue that these types of waste are necessary evils and that it is not possible to produce items without transport, or storage, or waiting. While this is almost always true we have to understand why we look at these operations as waste. The reason is that we need to look at these operations that do not add value as operations to be eliminated. Whenever any particular waste is eliminated or reduced, the organization becomes more efficient and more competitive. On the other hand, whenever one is aware that there is waste, there is always space for improvement by reducing it. Discovering waste can be seen as discovering treasures because whenever waste is recognize d, a potential space for improvement opens up.

Some renowned authors make the distinction between necessary waste and unnecessary waste. While being aware of the reason for this distinction it can also be argued that if all waste is assumed as unnecessary it will always be seen as potential spaces for improvement. No matter how difficult it is to eliminate an operation that does not add value, we can never lose sight of the fact that the elimination of that operation will result in a permanent benefit for the organization.

A typical example of this idea is the quality control operations. Quality control does not add value to products because it does not change their physical or chemical characteristics, but it is essential that this operation exists to ensure that products are produced within specification limits. If we look at this operation as a necessary waste, this operation will never be subject to study for elimination, but if we constantly look at it as a waste, there will always be the need to question its existence and to continuously seek a solution so that this control is no longer necessary.

Waste is not equally classified by all authors nor equally adopted in all organizations but the differences are not very relevant. We propose the use of the classification of waste published by Taiichi Ohno (Ohno, 1988), because he was the very much involved in the original waste classification during the development of the TPS. According to this classification, waste can be of the following types:

- Overproduction
- Waiting
- Transporting
- Too much machining (over-processing)
- Inventories
- Moving
- Making defective parts and products

To reinforce the negative impact of these wastes on organization performance some authors make a curious analogy calling them "the seven deadly wastes". Each one of these types of waste will be explained in next paragraphs.

Waste with overproduction - Overproduction is for many authors the worst waste of all because it is a direct generator of another very bad waste, inventories (Productivity Press Development Team, 1998). Producing more than is needed at the time is a very popular waste in the traditional way of thinking about production. Producing more than is necessary can be the consequence of 3 different classes of behavior: (1) producing more than is requested by the next process, (2) producing before the time requested by the next process and (3) producing at a rate higher than the demand rate of the next process. Some examples of typical causes of overproduction are as follows:

- Machine setup times for batch changeover are one of the most common causes of overproduction. Any one of us would argue that if the machine takes too long in its setup then the same item must be produced for quite a while to compensate the time spent in setting up the machine to produce that item. As the setup cost will be diluted over the items to be produced then the greater the number of items to be produced the lower the cost assigned to each one of them. The seemingly legitimate argument that it is not economically viable to keep changing the machine for a different item is very popular with production staff in traditional management thinking. What is usually overlooked, however, are the costs associated with the overstocking and long lead times resulting from that policy. Reducing machine setup times are the answer to this waste, using the classic tool, SMED[2]. An important contribution to promote this technique to pursue quick changeovers in Europe and America west was a book from Shigeo Shingo published in 1985 with the title "*A Revolution in Manufacturing: The SMED System*" (Shingo, 1985).

- Attitudes such as: "now that we have got our hands full, we can make some more", "we are sure we will need this later" or "let us produce more, just in case", which are very typical in our organizations, as reasons to generate overproduction. Producing more than it is necessary to finish that batch of raw material ("there is no point in leaving this small quantity"), to finish filling the box or to finish the shift ("now I will produce the same until the end of the shift"), are also common arguments that contribute to this type of waste called overproduction.
- Producing before the moment it is needed is also overproduction. Producing a batch ahead of schedule because the machine was available or because the machine setup was already performed for that type of item is waste in the class of overproduction. This behavior is not tolerated in world-class organizations. Toyota has not allowed it for several decades (Shingo, 1989).

With overproduction, more raw materials are consumed and more labor is used without any short-term (often never) return. Overproducing results in the need for more storage space, more capital tied up, more material handling and movements, as well as increasing production control problems. Living with overproduction, which can be translated into (and leads to) the constant occupation of equipment and people, giving managers the idea that everything is going well in production, hiding a huge amount of waste. The idea that nothing can be improved because all resources are always busy is a psychological posture that managers who are less conscious of waste can fall into for a long time.

---

[2] SMED (Single Minute Exchange of Die) is a technique for the reduction of changeover time that will be briefly presented in the annex.

Furthermore, as equipment and staff are always busy, at certain times the apparent need arises to buy more equipment and recruit more staff that would not be necessary if overproduction was not there.

Waste of waiting (people waiting) - This type of waste is mainly related to the loss of use of resources (equipment or people but mainly people) because they are waiting for something. It can be said that waiting can be classified in 4 types: people waiting for machines or materials, people waiting for people, machines waiting for people and machines waiting for machines or materials. This type of waste, like stock, is a symptom of inefficiencies in the production flow. However, unlike stock, it is viewed negatively by all managers, directors and production managers. No one with responsibility in production likes to see workers simply waiting for something to happen. This type of waste is frequently caused by capacity imbalances between processes, production scheduling problems, poor material management and machine breakdowns.

The Waiting type of waste can also include the time spent by operators monitoring automatic machines. It is very common to have operators simply looking at automatic loading and unloading machines, or during automatic production cycles, in order to prevent some problems that may occur either in the loading or unloading systems or in the process itself. This practice can be changed by eliminating the need for people monitoring machines, creating *Poka-Yoke*[3] type systems (see annex for more information) to prevent possible errors or creating automatic mechanisms to stop the equipment if something goes wrong. In the limit, when it is not possible to solve it this way, it is necessary to look for solutions in which the same operator monitors several machines at the same time.

Some experts say that it is a good practice to, when possible, turn all other types of waste into waiting to make the problems visible or bring problems to the surface. If you force stocks between workplaces to be limited to a minimum and this results in waiting type of waste, you are turning a problem that already existed in a hidden way into a visible problem. This is good because this problem easily starts to bother supervisors and managers and so something will have to be done to solve or minimize it.

Waste with transporting - The transport of materials, components and products in a productive system is as present as the process operations are on those materials. The Transport in a production system is closely linked to

[3] *Poka-Yoke* devices or systems, also known as Error-Proofing systems are systems or devices with the purpose of preventing errors. This tool will be briefly covered in the annex.

material flow, as there is always a flow of materials from suppliers to customers, through raw material warehouses, intermediate warehouses and finished goods warehouses. All the products produced need to be transported through all the production processes and this is a considerable effort in energy and manpower without resulting in any added value for the products.

The distances that materials have to travel are closely linked to the way workstations are arranged in the plant or factory, i.e. they are very much correlated to the existing layout in that plant or factory. The existing layout is often the result of production growth and the physical layout solutions that have been found to cope with that growth. It is also often the result of a misunderstanding of the objectives that should have been present in the design of the production system

The distances that materials have to travel have an impact on another type of waste, two of the most important types of waste, overproduction and inventory. The greater the distance the materials have to travel, the greater the natural and justified tendency to transport larger quantities at a time to reduce the number of times the distance is travelled. Now, when you want to gather enough quantity to justify the transport, you are creating overproduction and inventory of goods in progress.

Waste with too much machining (or also called over-processing) - Not all processing is usually necessary or carried out in the most effective and efficient way. We often find that the wrong working methods are continuously being carried out, poorly functioning equipment is being used, or inappropriate tools are being used in production, resulting in several efficiency losses. It also happens that sometimes a set of operations carried out at different workstations could be combined at one workstation and all those operations completed in one process. In other cases, a hole that is no longer required is still drilled in a part simply because the work instructions for that operation have not been updated and no one on site has been informed of the change. These situations exist in most organizations without their managers being aware of them. It is very natural that both managers and workers fail to see some things that are wrong. In some cases, the workers themselves do not try to improve the processes because they do not find in their supervisors and manager, maybe for lack of time, the necessary attention to help solve those problems. The truth is that much of the over-processing waste remains part of the normal day-to-day routine unless someone takes the initiative to tackle it.

Waste with Inventory (or Material on Hold as referred by Euclides Coimbra (Coimbra, 2009)) - Inventories are the accumulation of materials, articles, components or products at any stage of the production process. Inventories are usually justified for one or more than one of the following reasons:

- Imbalances between capacities of different processes in a production sequence. If an upstream process has a higher throughput than the downstream process, it is natural that stock is created. The opposite can also be true. There are cases where one part of a process works two shifts a day to satisfy the next process that only works one shift a day, in which case it is obviously necessary to keep stock.
- Equipment temporarily unavailable. If a piece of equipment is unavailable for some time, it is natural that the upstream process does not stop producing, creating stock.
- Suppliers' missed deadlines. If suppliers are more likely to fain in meeting their deadlines, it is natural for the organization in question to keep materials in stock to avoid shortages in the cases the supplier do not meet the defined deadlines.
- Defects. It is necessary to have good products in stock to be able to replace defective products that may be produced.
- Long lead times. Long lead times naturally lead to the production of stock items (see overproduction)

People in general in organizations, regardless of the department they work in, don't worry too much about having too much of something, they only get very worried when something is missing. This is the nature of human thinking and for that reason inventory naturally exist in higher levels than really needed in most organizations. Inventories put people at ease but are the cause of huge losses in productivity and speed of response to market. Inventories are a particularly bad kind of waste because they help to hide a large number of problems in production units and plants (Productivity Press Development Team, 1998).

The existence of inventories throughout the production processes causes other types of waste such as transport, defects, handling and movement. The reasons that may lead to the need for transport are related to the fact that it is often necessary to move stocks to an intermediate storage area and then pick them up and transport them to next processes. On the other hand, the accumulation of stock in the manufacturing area frequently forces transport equipment (trolleys or forklifts) to bypass these stocks. It is also common for stored materials to deteriorate over time, to deteriorate due to poor packaging, or, what sometimes happens, because they are "run over" by forklifts or other transporters. Finally, stocks generate the need for movements which would not be necessary if they did not exist. These movements concern loading and unloading into storage areas and sometimes the need to move those materials. Sometimes there is also the need to move materials from their storage areas in search of other materials.

Waste with moving - Waste with moving refers to all movements of people or machines that do not result in any added value in the products. Examples of this type waste are:

- People searching for materials, drawings or documents.
- People looking for tools in drawers, on shelves or around the factory.
- People looking for people (foreman, boss or colleague) to gather information.
- People going to the warehouse to get missing components for production.
- People walking from one machine to another or from one section to another.
- People taking messages or delivering documents.

These movements are a normal part of daily work and are not intentional, in fact they are easily justified, but they are the result of inappropriate equipment layout, lack of attention to ergonomic aspects, poor organization of workstations, lack of cleanliness, inconsistent working methods, etc.

Waste of Making defective parts and products (sometimes just called defects) – This waste is generated from the production of defective items in the production process. It does not only concern products that do not pass quality control at the end of production but all the defective items produced in every step of the production process. It is often the case that some defects caused in some processes are repaired and eventually follow the normal route to becoming defect-free end products. In these cases the waste also existed and had a negative impact on performance. Making defective parts and products, among other causes, results from:

- Manufacturing errors.
- Installation errors.
- Design errors.
- Requirement interpretation errors.

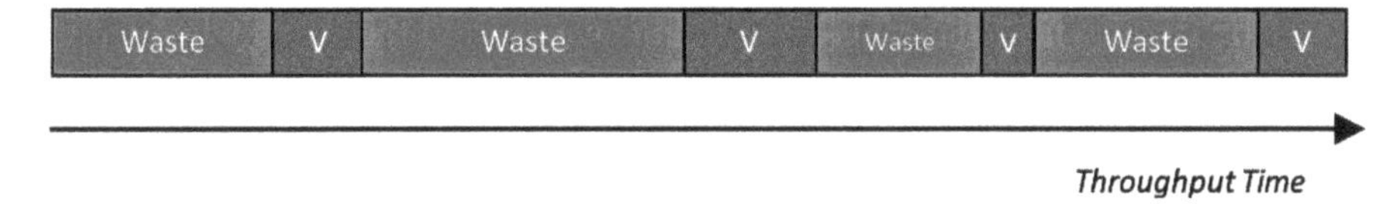

Figure 2.5 Generic representation of a value chain.

- Communication errors.
- Lack of skills or training.
- Incorrect transport or packaging.
- Incorrect material use.

Principle 2 - "Identify the Value Stream" - This means identifying all the processes involved in the process of producing the product, not only those that add value (value from the customer's point of view) but also those that do not add value. The latter are the so-called wastes and should be eliminated or at least minimized. The most popular tool to analyze the value stream is probably the tool called Value Steam Mapping[4] which consists of representing in a kind of flow chart the connections with mains suppliers and clients, the production planning and control main communication channels, all main stages of the production process as well as some details of each stage in order to identify main sources of waste and bottlenecks in the flow.

Figure 2.5 shows the representation of a simplified example of the value stream for a hypothetic and generic product or a family of products. In this representation the reader can see in a time scale the sequence with 4 processes where value is added (represented with a "V"), and four periods where time is spent without any value being added to the items (represented with "Waste"). These "waste" periods include mainly transport and handling, materials waiting (inventory), rework and eventually quality control steps. Typically, inventory plays the major part of waste in the timeline. The concept of Throughput Time is a very important concept and indicator that must be understood when analyzing value stream and the concept flow. Throughput Time can be defined in general terms as the amount of time required for a product to go through a manufacturing process, being converted from raw materials into finished products. For the case expressed in Figure 2.5 the throughput time is the sum

[4] Value Stream Mapping (VSM) (Rother & Shook, 1999)(Jones & Womack, 2002) will be briefly presented in the annex.

of all the times when materials are waiting (waste) and the times when value is being added with operations to those materials.

It is often referred in several publications, such as Beecroft, Duffy, & Moran (2003) and Productivity Press Development Team (1998) to a round figure of 95% for the typical percentage of Throughput Time that is spent without any value being added to the products (see Figure 2.5). In other words, we can say that 95% of the time that an article is within a production system that article is standing still waiting for something to happen to it.

This 95% value is referred to and used as a reference but in fact the most important aspect of this round value lies in the transmitted idea that the overwhelming part of the time of the articles in production units is spent waiting (as inventory) for the next process. In the various evaluations we have performed in dozens of organizations in the last decade, this value is normally higher. It is also important to note that everything depends on where we start counting time and where we stop. If we consider that the throughput time should start counting when the materials arrive at the raw materials warehouse and only stop counting when the products are shipped to the clients, it is natural that this percentage is higher than if we count the time from when we pick up the materials from the warehouse until the products are sent to the finished product warehouse. When evaluating the throughput time the analyst must clearly establish the boundaries of the systems with managers so everyone knows what the real meaning of the results. The throughput time can also be evaluated to specific parts of the production process, such as production lines of specific processes steps.

Principle 3 - "Flow" – This principle consists in making sure that the products being manufactured stop waiting as little as possible, i.e. that products should always being transformed in an operation/process or being transported (as little as possible) to the next operation/process. Making an analogy with fluid mechanics, fluidity is maximum when all particles of the fluid move easily in the same direction. This is what is desired in production, that all materials, all components, all parts move towards the customers with as little downtime as possible. Whenever items stand still waiting for something to happen, we do not have fluidity. In an ideal situation the items would only be stopped when performing required value adding operations, otherwise they should be moving to the next process. This ideal situation is only theoretical, represents a vision, but the objective of such vision is that we should always try to find solutions to improve fluidity.

A very effective indicator to measure the fluidity of production is the Value Added Ratio (VAR) (Shannon, 1997) which is evaluated by dividing the total time of value added operations of a product of family of products by the average throughput time of that product or family of products. The Throughput Time of a product in a production unit is the time between the moment when a unit of that product (in the form of raw material, for example) enters the production unit and when it is ready and shipped from that same production unit to the client (even if internal client of that production unit).

$$VAR = \frac{\sum Processing\ Times}{Throughput\ Time}$$

The value of this indicator is typically very low and very often even below 1%. Although this value varies greatly between different types of industry and although it is not easy to establish benchmark values with any reliability, the higher the value, the better the flow of materials in production and consequently the more competitive the organization is. To have a clearer idea of the meaning of this indicator, if we consider a value of VAR equal to 1% then it means that 99% of the time that materials spend inside the organization, these materials are stopped waiting for something to happen. The reader must be aware that it is the reality in most of our organizations. The 5% value for this ratio is used, as mentioned earlier, as a reference for world-class organizations, that is, very competitive organizations with high performance.

Principle 4 - "Pull" – The meaning of this principle is that ideally an operation/process should only be executed in a product when the downstream operation/process needs it, with delivery to the customer being the last of these operations/processes (i.e. the customer "pulls" production). Adjusting the flow of all items going forward in production to demand depicts this fourth principle of Lean philosophy, dubbed Pull Production. The more that decisions to move items through the various steps of the entire process are dictated by demand, the better the production performance will be. Creating flow only makes sense if demand dictates how items flow through the production process, otherwise products would be accumulating in the finished products warehouse. There are many obstacles to achieving this behavior, but the idea is to always move towards being the market the driver in pulling production.

Some production processes make production pulled by demand and production flow very difficult. Some examples are the following:

- Processes with long setup times. In this type of processes, such as stamping on presses, or injection of plastics, the quantity to be produced can be higher than the quantity pulled by downstream processes in a particular moment.

- Processes that require the grouping of components from various different products to optimize the process, such as cutting sheet metal or fabric. To better use material such as in textile cutting processes, sometimes is better to cut components from different finished products in order to reduce scrap of textile material. Some of the components are not needed, in that moment, in downstream processes.

This topic of pull production will be discussed in more detail in chapter 5.

Principle 5 - "Perfection" – When the previous four principles are applied people realize that there is no end to the process of reducing waste and improving pull flow. This principle reflects the idea that the other principles must be continuously pursued in cycles. That is why some experts also call this principle as "Continuous Improvement". This principle is of enormous importance and Toyota clearly gives it great emphasis, as shown in Figure 2.1. Sustaining continuous improvement, or pursue perfection, requires major transformations, even disruptive ones, in the structure and culture of organizations. It is in this fifth principle of the Lean philosophy that one can somehow frame the second basic concept of TPS, suggested in 1977 (Sugimori et al., 1977) and much of what is related to human factors, management and culture. According to these authors, making full use of employees' capabilities may be regarded as equivalent to treating them as human beings and with consideration. This idea can be divided into the following three main aspects:

- Elimination of waste due to movements without added value. The use of people in activities without added value that can be eliminated is, in a way, a form of disrespect towards them.

- Permanent consideration for the safety of employees. It must be ensured at all costs that employees are always safe from any risk of accident. Guaranteeing the safety of all employees is unquestionable. At the same level of safety, one can also refer to not exposing employees to environments that may cause them long-term problems such as the development of occupational diseases.

- Exposure of employees' capabilities by giving them more responsibility and authority. Giving more responsibility and authority is absolutely central as it promotes the use of human potential.

This last aspect is also referred to by Jeffrey Liker in his book entitled "Toyota Way: 14 Management Principles from the World's Greatest Manufacturer" (Liker, 2004) as the eighth waste - the non-utilization of human capital or the unused creativity of employees. This waste was described by this great author as being "wasting time, ideas, skills, improvements and learning opportunities by not involving or listening to your employees". These are the reasons why the principle of continuous improvement (or pursuit of perfection) can be considered as closely linked to respect for employees, since no other principle of the Lean philosophy is exclusively dedicated to this aspect.

2.3 Principles of Toyota Way

At the beginning of the 21st century, the social sciences' side of organizations started to gain more and more recognition in many organizations around the world. One of the organizations that clearly include that invisible side in the form of principles was again Toyota by creating the Toyota Way excellence model. The Toyota Way is one of the models of excellence whose principles very effectively cover the entire spectrum of the socio-technical nature of organizations (Liker, 2004). The principles with grey background in Table 2.1 are principles more linked to the continuous improvement side of the Toyota Way while the other ones are more linked to the Respect for People side.

The *Toyota Way* model was formally established at Toyota in 2001 (Toyota_Motor_Corporation, 2012). One of the reasons for the adoption of this new model is the fact that the existing *Toyota Production System* was mainly focused to the production area. The *Toyota Way* is a more comprehensive model that extended and adapts the existing TPS concepts to all other sectors of the organization. Just as in Toyota, this extension to areas other than production has also occurred in various organizations around the world. Thus, with the designation Lean <something>, the most varied implementations appeared, for example, in hospitals (*Lean Hospitals*), in construction (*Lean Construction*), in accounting (*Lean Accounting*), in administrative environments (*Lean Office*), in service environments (*Lean Services*), in academia (*Lean Teaching*), as well as in many other areas.

This *Toyota Way* is referred to on the Toyota Europe website (Toyota_Europe, 2019) as being based on two pillars (already mentioned above): Continuous Improvement and Respect for People (includes respect and teamwork). Toyota Way clearly values the respect for people pillar expressed for instance by highlighting the labor-management relations based on mutual trust and respect (Toyota_Motor_Corporation, 2009). It is curious the proximity

Table 2.1 Toyota way 14 principles liker (2014).

Section	Principle
Section 1 Long-term philosophy	Principle 1 – *"Base your management decisions on a long-term philosophy, even at the expense of short-term financial goals"*
Section 2 The right process will produce the right results	Principle 2 – *"Create a continuous process flow to bring problems to the surface"* Principle 3 – *"Use 'pull' systems to avoid overproduction"* Principle 4 – *"Level out the workload (Heijunka)"* Principle 5 – *"Build a culture of stopping to fix problems, to get quality right the first time"* Principle 6 – *"Standardized tasks are the foundation for Continuous Improvement and employee empowerment"* Principle 7 – *"Use visual controls so no problems are hidden"* Principle 8 – *"Use only reliable, thoroughly tested technology that serves your people and process"*
Section 3 Add value to the organization by developing your staff and partners	Principle 9 – *"Grow leaders who thoroughly understand the work, live the philosophy, and teach it to others"* Principle 10 – *"Develop exceptional people and teams who follow your company's philosophy"* Principle 11 – *"Respect your extended network of partners and suppliers by challenging them and helping them improve"*
Section 4 Continuously solving root problems drives organizational learning	Principle 12 – *"Go and see for yourself to thoroughly understand the situation"* Principle 13 – *"Make decisions slowly by consensus, thoroughly considering all options; implement decisions rapidly"* Principle 14 – *"Become a learning organization through relentless reflection and continuous improvement"*

with the concepts that were presented few decades earlier (Sugimori et al., 1977)).

The *Toyota Way* is broken down into 14 principles that were published by Jeffrey Liker (Liker, 2004) and are grouped into the following four sections:

Section 1. Long-term philosophy.
Section 2. The right process will produce the right results.
Section 3. Add value to the organization by developing your staff and partners.
Section 4. Continuously solving root problems drives organizational learning.

The 14 principles of *Toyota Way* are presented in Table 2.1, grouped by section, where one can observe the great emphasis given to the human factors or social sciences side of the socio-technical nature of organizations. In addition to the whole of section 3 which is dedicated to this human component, we also have principles 13 and 14 in section 4.

Since some of the principles may not be completely self-explanatory and for that reason below we will provide a short clarification of each of them.

Principle 1 – *"Base your management decisions on a long-term philosophy, even at the expense of short-term financial goals"* - The existence of a long-term philosophy is like the existence of a "north" to simplify decision making. Let us use a simple example to show the scope of this principle. Imagine that the long-term philosophy includes concern for the quality of life, job satisfaction and health of employees. Now imagine also that there are two investment alternatives to meet a certain need, alternative "A" is more expensive but better for the health and comfort of the employees and alternative "B" is cheaper but worse for the health and comfort of the employees. In this case the long-term philosophy prevails over short-term financial goals, deciding for alternative "A".

Principle 2 – *"Create a continuous process flow to bring problems to the surface"* - The principle of continuous flow has been mentioned before as a principles of Lean Thinking and will also be presented in more detail in chapter 5 but we will clarify here the issue of the effect of flow on problem identification. Instead of hiding the problems as usual in our organizations, the idea is to have all problems clearly exposed for all people to see. The reduction in the quantity of work in progress, as a result of the continuous process, means that problems such as breakdowns, delays and defects are quickly noticed throughout the system because the "shock absorbers" for these problems, "stocks", exist in very small quantities. Imagine a line with

10 work stations where there is only one item between each work station. Now imagine that one of the stations has stopped producing due to a breakdown. In this case the whole line with 10 work stations quickly comes to a halt and it is very clear to everyone involved that there was a problem. If the inventory between jobs was a large quantity of units, there might be time to resolve the fault before the whole line stopped. Note that the breakdown problem may go unnoticed when there is a lot of inventory but it does not go unnoticed when there is little inventory (continuous flow). This is bringing the problems to the surface.

Principle 3 – *"Use 'pull' systems to avoid overproduction"* - Overproduction is one of the 7 classic types of production waste that were defined at Toyota and mentioned earlier in this chapter. Many are the publications that describe these 7 types of waste but the one that seems to us to be the most assertive is the one by Euclides Coimbra in his book "Total Flow Management" (Coimbra, 2009). Overproduction, or "too much production" as coined by this author, is all the quantity produced more than requested by the client (the client can be the next process), or all the quantity produced before being requested by the client (or next process). This type of waste is difficult for many people to understand because people frequently think that having too much product in stock is good. The idea is that if there is a requested for it there is no need to produce it because it already exists. Pulled systems are systems or mechanisms that prevent overproduction. These mechanisms do not allow upstream processes to produce more than the quantity required, or before it is requested, by downstream processes. Some of these mechanisms will be presented in chapter 5.

Principle 4 – *"Level out the workload (heijunka)"* - First of all one has to produce only at the rhythm or pace of demand. With that in mind the idea of "*heijunka*" is to produce small quantities of all types of products every day or every shift in order to meet the corresponding demand during those periods. If there is demand for 10 types of products then you should try to produce every day (or every shift) the quantities that meet the demand for each of them. In this way you only have to hold in stock quantities for one day or shift. It is a way of helping to maintain a continuous and pull flow. Seen from another point of view it is also a way to reduce variability (Mura).

Principle 5 – *"Build a culture of stopping to fix problems, to get quality right the first time"* - This was also a principle that contradicted the way of thinking that existed in production and is still for many, difficult to understand. To clarify the concept let us imagine a car assembly line. Now imagine that a quality problem was detected in one of the cars on a section of the line. The traditional approach, and probably the most accepted common sense, is to remove the car from the line and keep the line running. The removed car will then be repaired by specific employees for that function. I imagine that for most people this way of approaching the problem is nothing more than “common sense”. Does it make any sense to stop the whole line and leave dozens of employees waiting for the problem to be solved? This is exactly what Toyota thought should be done. All employees on the line are instructed to stop the line when a problem occurs and the line only starts producing when the problem has been solved.

During a visit we had with students from our University to the Bosch-Vulcano industrial plant in Aveiro, an important piece of packaging equipment broke down and to the students’ were amazed with what happened as a consequence. The entire factory stopped producing. Several hundred workers were kept waiting, without producing anything for a couple of hours, until the packaging equipment started working again. The decision to stop production is better than letting end products accumulate without being packaged. The impact of this decision is that all the focus is now on finding the root cause of the problem so that another breakdown doesn’t happen. Although it was not a quality problem, the message is the same, if they had continued the production they would have hidden the problem and they would have had to deal with large amounts of unplanned operations with huge difficulties in space management, people management and equipment use.

Principle 6 – *“Standardized tasks are the foundation for Continuous Improvement and employee empowerment”* - Any task must be carried out in a standard way and there should be no improvisation. If someone has a better way of carrying out a task then it must be checked to see if it is better and if it is, that way becomes the new standard. Carrying out tasks in the same way over and over again ensures that the result is always the same, in terms of quality, processing time and safety. One example where the concept of standardized work is taken very seriously is in aviation. The pilot and co-pilot always carry out the same routines because in that way accidents and incidents are minimized and the safety of the passengers is better guaranteed. Even in the event of a malfunction during the flight, pilots will have to follow pre-established standard instructions.

Principle 7 – *“Use visual controls so no problems are hidden.”* - Transparent information and effective communication are important aspects of performance and people’s well-being. Any simple visual solution that clarifies what is happening in production is a huge help for employees and managers. Take the classic example of "Andon"[5] systems applied to machines or lines to indicate their status or painted areas on the wall to indicate upper stock limits. When *Andons* are implemented, by simply looking, managers and workers know what problems exist, if machine are working properly or not, if the performance is above or below the objective and so on.

Principle 8 - "*Use only reliable and thoroughly tested technology that serves your people and processes*" - This principle is a very interesting and maybe dubious since new technologies can be very promising. This is very conservative and cautious principle, being effective in helping to avoid the temptation of adopting promising new equipment without full assurance that it is reliable and fits many other principles such as production fluidity, quality and safety.

Principle 9 - " *Grow leaders who thoroughly understand the work, live the philosophy, and teach it to others* " - The entire leadership chain must be developed so that their daily behavior and actions are aligned with the organization's philosophy. Personal development of employees must be assumed as part of the leadership's responsibilities. In addition, leaders should have in-depth knowledge of the shop floor (*Gemba*), the equipment and the personalities and needs of the employees and should not fall into the temptation to make decisions based on reports. Leads must get information and data mainly by observing the processes and talking to operators.

Principle 10 - "*Develop exceptional people and teams that follow your company's philosophy*" - The idea that people are resources with infinite capacity must be cultivated in the organization. Equally relevant is that the awareness that synergies achieved with true teamwork must be present in the leadership. The right teams in the right context can achieve extraordinary achievements and therefore teamwork should be strengthened. Leaders must continuously seek to contribute so that employees grow as professionals and also as people.

Principle 11 - "*Respect your network of partners and suppliers by challenging them and helping them improve*" - Both this and the previous principle fall under the umbrella of "respect for people" although in this case it also includes creating robust and lasting partner relationships with partner organizations and suppliers. This principle is another case of a paradigm shift from traditional thinking. Instead of trying to take advantage of competition

---

[5] Andon are systems that notify employees and managers in case of disruptions in production processes. A briefly description of such systems can be found in the annex.

between suppliers to get better prices or better conditions the idea is to choose only a few suppliers and create a deeper connection with them. This deeper relationship gives on one hand guarantees of continuity to the supplier and also promotes the dissemination of concepts and practices of excellence, helping partners to grow and evolve in excellence in organizations and in continuous improvement. The idea is that these lasting partnerships result in long-term gains for all.

Principle 12 - "*Go and see for yourself to thoroughly understand the situation (Genchi Genbutsu)*" - This principle is also known as "going to the Gemba". Many traditional managers make decisions based on reports and/or third-party reports and this is a practice with a huge risk of giving rise to errors of judgement and consequently errors in decision making. Decision makers should, according to this principle, go to the field frequently in order to get to know the true reality on the ground. It is surprising how by talking to people in the field, observing the equipment and the operation, one can get a much more robust idea so that decisions can be taken much more effectively.

Principle 13 - "*Make decisions slowly by consensus, thoroughly considering all options; implement decisions quickly (Nemawashi)*" - This is an interesting question that deserves some reflection. Let's divide this principle into two parts, the first part being about the decision making process and the second part being about rapid implementation. Regarding the first part, a very commonly used approach in our organizations is decision making based on majority but there are also cases where the decision is made by the leader and then "accepted" by the others. These decision making processes although relatively effective for decision making may not work very well after implementation as some of the participants may not feel very committed participate in to the decided solution. The idea of this principle is that the decision process should involve everyone so that everyone feels committed to the decision taken.

Also related to this principle, there is a decision-making system also practiced in Japan which is called "*Ringi Technique*"[6] (Sagi, 2015) which allows consensus between peers to be achieved in order to avoid face-to-face confrontation. In Eastern cultures such as the Japanese, the open confrontation of ideas and opinions as we do in the West is not very comfortable for the participants. For that reason this *Ringi technique* was developed to avoid that discomfort. It all starts with a document, "*ringisho*", describing the problem and the solution anonymously. Then this document is circulated to all the

[6] This technique to help in finding consensus is referred in the annex.

members of the group one by one. Each member puts their own stamp on it and adds their own comments and suggestions. When the author receives the document with the comments and suggestions, he rewrites the document to accommodate these comments and suggestions. The process continues until the comments no longer appear, in which case the decision is made by consensus. This is a slow process but can generate good consensus solutions among all group members without the need for confrontation.

A similar process is used by Frank Devine in his method called "*Rapid Mass Engagement*" (Devine & Bicheno, 2020) which aims to develop commitment of all employees and managers to the goals of the organization. In one of the last stages of the process there is an event called "consensus day" where many employee representatives and the management team take part and the aim is to reach a general consensus on rules of behavior and action plans. This event has no time limit and if necessary continues the next day until everyone agrees on the decision.

Regarding the second part of this principle - rapid implementation - it happens that in many of our organizations, although decision making is relatively simple and quick, implementation can take a long time or simply never be implemented. Moreover, in our culture, unfortunately, it also happens that a decision is taken about a certain rule to be implemented but then everyone fails to apply it. The reader will certainly find examples of these in his or her context.

Principle 14 - " *Become a learning organization through relentless reflection (Hansei) and continuous improvement (Kaizen)*" - This principle includes the concept of "*Learning Organizations*" which became very popular after being presented by Peter Senge in 1990 in his book "*The fifth discipline: the art and practice of the learning organization*". A second edition of the same book was published in 2006 (Senge, 2006). Learning organizations are organizations where there is a culture that includes, for example: open sharing of information and ideas, encouragement of critical thinking, openness to learning from mistakes, knowledge building using PDCA cycles, peer review, encouragement of personal growth for all, seeking customer feedback, getting customer feedback on products under development as early as possible, and sharing a common vision. In summary it can be concluded that this principle reflects the essence of continuous improvement and everyone's growth towards mastery.

2.4 Principles of the Shingo Model

The *Shingo Model* is another relevant reference that could not be ignored in this book and that also puts clear emphasis on the human aspects, just like in the *Toyota Way*. The Shingo Model (Plenert, 2017) is naturally connected to the *Shingo Prize* created by the *Shingo Institute* in 1988, curiously at the same time that the term "Lean" was being coined by the MIT group led by James Womack previously mentioned (Krafcik, 1988).

> The Shingo Institute was created in 1988 at Utah State University in the United States of America following an honorary doctorate awarded to Shigeo Shingo himself. Shigeo Shingo was a great consultant with a long career in the Japanese automotive industry and also contributed, with Taiichi Ohno, with several theories and methodologies to the development of the *Toyota Production System*. He wrote several important books including "*A Study of the Toyota Production System: From an Industrial Engineering Viewpoint (Produce What Is Needed, When It's Needed)*" first published in Japan in 1981 and then published in the USA in 1989 (Shingo, 1989). The institute was named in his honor.

Shingo Model started to be developed in 1988 to support the Shingo Prize, awarding the first organization in 1989 located in Berkeley in the USA, called Globe Metallurgical (Shingo Institute, 2020). Since then, it has been awarded to numerous organizations that meet the requirements set for this important excellence award. The Shingo award is based on a set of criteria with the aim of assessing how "*Lean*" an organization is, or, put another perhaps more correct way, how close to the Toyota way of working an organization is. The first version of the Shingo model, also referred as "1st Assessment Model" was established in 1993 (Shingo Institute, 2106). Very little emphasis was given in that version to the human side of organizations and no reference was given to continuous improvement concept. A new Shingo Model was released in 2008 (Shingo Institute, 2106) with emphasis on principles and culture where clear relevance was given to continuous process improvement, assigning a set of principles to that dimension. The actual version of the Shingo Model (Miller, 2018) is very much an enhancement of that new Shingo Model. In the point of view of scientific publications the first article found in Scopus database referring the Shingo Model was published in 2014 (Bhullar et al., 2014). In that article, the authors refer the Shingo Institute website in 2012 as the source of those principles. The ten guiding principles are categorized into three dimensions – Cultural Enablers, Continuous Improvement, Enterprise Alignment, and Results, as shown in Figure 2.6. The first dimension of the guiding principles lies on the Culture Enablers

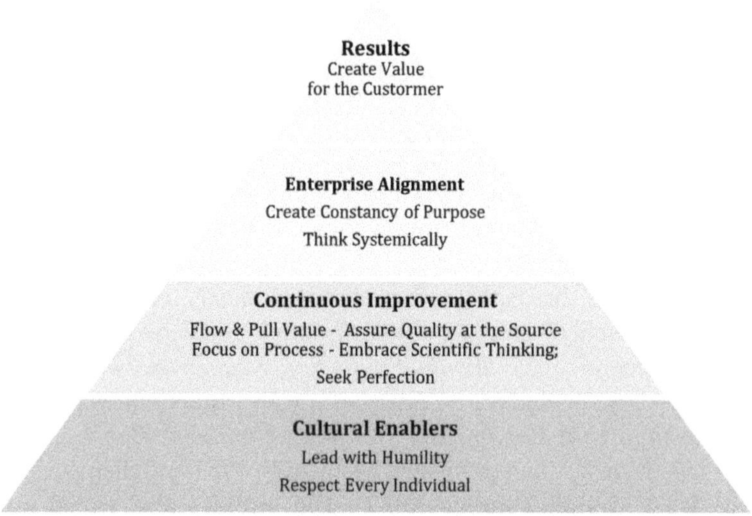

Results
Create Value
for the Custormer

Enterprise Alignment
Create Constancy of Purpose
Think Systemically

Continuous Improvement
Flow & Pull Value - Assure Quality at the Source
Focus on Process - Embrace Scientific Thinking;
Seek Perfection

Cultural Enablers
Lead with Humility
Respect Every Individual

Figure 2.6 Shingo model guiding principles.

principles of respect for people and lead with humility, and they are at the bottom of the pyramid because they concentrate on the foundation of an organization: the people. This class refers to the type of behaviors required in order to effectively accommodate all the other principles. The second dimension of the guiding principles pyramid – Continuous Improvement – refers to the principles related to the production processes focus (Pull, Flow, Quality, and Scientific Thinking) and its improvement seeking perfection. The "Enterprise Alignment" dimension refers to the formal vision and purpose of the entire organization. The "Results" dimension refers to the importance of creating value to the customers as a tool to getting great results in terms of consistently delivering ideal results to all stakeholders

The principles of this model can be assigned to each one of the sides, technical and social, in a relatively easy way. Some of the principles in the class "Continuous Improvement" can be assigned to the technical side while the principles in the other classes, "Cultural Enablers" and "Enterprise Alignment" can be assigned to the social sciences side.

If we compare these principles with the principles of *Lean thinking*, it is easy to identify the differences. The Shingo principles are much more comprehensive than the Lean principles and probably reflect the reality of Toyota's philosophy and culture in more detail, since they put a lot of emphasis on the less visible aspects, which are associated with human involvement and culture.

The first two principles are clearly dedicated to the respect for the individual side, referred by (Sugimori et al., 1977), while principles 8 and 9 are more related with the management aspects. Interestingly, five principles are dedicated to the class of continuous improvement and include almost all technical aspects and behaviors with direct influence on the production process. Similar to the clarification which was done for each principle of the *Toyota Way*, a brief clarification of each principle of the *Shingo Model* will also be presented below.

Principle 1 - "*Respect Every Individual*" - This principle is one of the main concepts of the Toyota way, along with the concept of continuous improvement (see Figure 2.4). It is not only about being polite to all individuals, but also about allowing each individual in the organization to make full use of his or her capabilities and to grow as a person with increased self-esteem and a sense of self-realization.

Principle 2 - "*Lead with Humility*" - Humility is a very precious ingredient for learning and improvement to happen effectively. Any leader who is always open to learning from his subordinates is a leader who will be successful with his or her team. The humility of the leader allows each one led to feel that their contribution can be useful for the whole team. A leader who is always looking for contributions from his subordinates and listening to them attentively and actively, will be creating an environment where the subordinates feel available to contribute openly with their creative skills.

Principle 3 - "*Seek Perfection*" - This principle is equivalent to the fifth principle of the Lean philosophy.

Principle 4 - "*Embrace Scientific Thinking*" - Instead of simply using intuition, it is important to adopt scientific thinking using the scientific method embodied in the PDCA cycles (see in Annex). These cycles include experimentation, direct observation and learning. Systematic exploration of new ideas, with openness to unsuccessful experiments, allows for increasingly accurate understanding of reality.

Principle 5 - "*Focus on the Process*" - The main idea of this principle is that results are always the consequence of the process. In our traditional organizations it is very common to blame people for problems. The truth is that inefficient processes or processes with enormous variability will always produce mediocre results. It is important that processes are well designed and continuously improved so that people can carry out tasks with greater added value.

Principle 6 - "*Assure Quality at the Source*" - This principle is equivalent to principle 5 from the *Toyota Way*. Quality is not separated from process. We shouldn't have one producing and the other assuring quality. Whoever produces must produce well the first time and whenever there are problems, the root cause must be found and a countermeasure implemented so the problem doesn't arise again.

Principle 7 - "*Improve Flow and Pull Production*" - This principle is equivalent to the sum of principles 2 and 3 from the *Toyota Way* and the sum of principles 3 and 4 of the *Lean Philosophy*. Making value flow and pull production was the most copied part of Toyota in the West during the late 20th century. Chapter 5 of this book is dedicated to this subject.

Principle 8 - "*Think Systemically*" - To think systemically is to always think of the whole and not just the part. Improving one step of the whole process may not mean improving the whole and so it is important that everyone understands the whole. A practical example that is often used in *Lean* approaches is the overview based on process mapping using VSM (*Value Steam Mapping*) (see Annex). This analysis focuses on the whole process and how each step in the whole process can influence overall performance. Pull and flow systems are ways of dealing with the process as a whole. Besides the aspects related to the physical part of the production it is important not to forget the communication between the actors of different steps of the process as good communication and sharing allows a better view of the whole. There are huge gains whenever the downstream process and the upstream process share their problems because many of them can have useful solutions for both.

Principle 9 - "*Create Constancy of Purpose*" - This principle can be considered equivalent to principle 1 of the Toyota Way. The vision and purpose of the organization is known and followed in an aligned way by all employees and managers.

Principle 10 - "*Create Value for the Customer*" - This principle is equivalent to the first principle of the Lean Philosophy.

2.5 Comparison of the Principles of the 3 Models

If we compare the principles defined for the *Lean philosophy* with those defined by the *Toyota Way* and with those defined by the Shingo Model, we can make some considerations. First, both the *Shingo Model* and the *Toyota Way* put much more emphasis on the human aspects and on deliberately building a specific type of organizational culture than the *Lean philosophy*. Both the *Shingo Model* and the *Toyota Way* clearly mention the aspects of culture and respect for the individual, but that is not mentioned in *Lean philosophy*. However, this does not mean that *Lean* does not have these concerns. Of course it does, since *Lean* is just a term that was created to describe the way Toyota organizes and manages itself. Perhaps the *Lean* principles, which have the advantage of being simple, have failed to address some important aspects. Nevertheless, there are, in our opinion, some aspects which, when expressed in prominent places, bring another consistency to the model. Some critical examples of such aspects appearing in the *Toyota Way* which can be highlighted are the already mentioned principles 1, 14, 10, 13 and 9, whose description is:

- Base your management decisions on a long-term philosophy, even at the expense of short-term financial targets;
- Become an organization that learns through relentless reflection (Hansei) and continuous improvement (Kaizen);
- Develop exceptional people and teams who follow your company philosophy:
- Make decisions slowly by consensus, thoroughly considering all options; Implement decisions quickly (Nemawashi);
- Create leaders who fully understand the work, live the philosophy and teach it to others.

Following the same logic, some of the aspects expressed in the *Shingo Model*, and which can be highlighted, are the principles 1, 2, 4, 8 and 9, whose description is:

- Respect every individual;
- Lead with humility;
- Embrace scientific thinking;
- Think in a Systemic way;
- Create Constancy of Purpose.

This less technical side of some of the principles present in both the *Shingo Model* and the *Toyota Way*, which we might associate more with the social sciences, is often overlooked by managers. The more direct, and perhaps more naïve, approach of managers and leaders is focused on the application of techniques and tools and the results that can be obtained. Many are the organizations that survive and compete in the marketplace resorting only to this way of being. True success is achieved when the organization focuses on a higher purpose and invests in principles, behaviors and building a culture centered on learning and developing people (Figure 2.7).

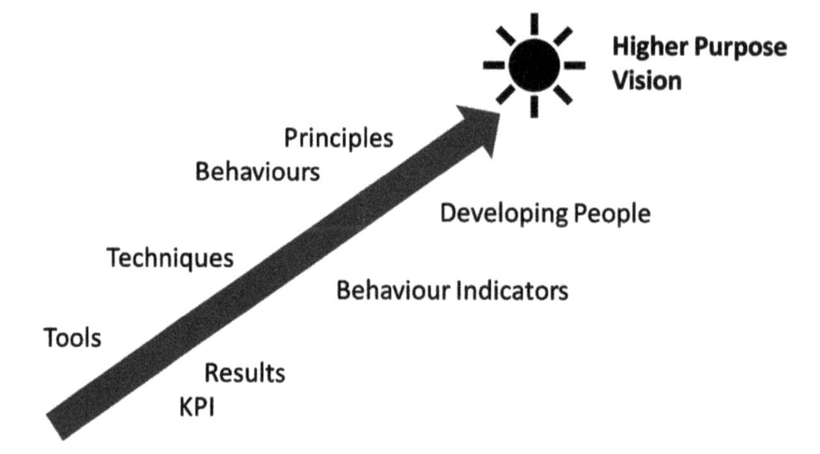

Figure 2.7 The role of techniques, principles and behaviors in the road to the higher purpose.

Table 2.2 Shingo model and Toyota way classes (dimensions or sections).

Toyota Way Classes	Shingo Model Classes
Long term philosophy	Enterprise alignment (purpose)
Add value to the organization by developing your people and partners	Cultural enablers (people)
The right process will produce the right results	Continuous improvement (processes)
Continuously solving root problems drives organizational learning	
	Results (stakeholders)

Analyzing the principles inherent to each of the three models we can detect some differences and also some similarities. The first major difference is the number of principles. While the *Lean philosophy* proposes only 5 principles, the *Shingo Model* proposes 10 and the *Toyota Way* proposes 14. Another important difference is that both the *Shingo Model* and the *Toyota Way* give much more importance to human aspects and organization culture than the *Lean philosophy*. Another difference between the *Lean* proposal and the other two is the grouping of principles in classes. Both the Shingo model and the *Toyota Way* group their principles into four classes (Table 2.2). As it can be inferred from the same table, one can even say that there is some equivalence between the first two classes in the two models.

For the other two classes, the equivalence is not so striking and it appears that in the *Toyota Way* there is more emphasis on the more technical principles of production. This is shown by the exclusive assignment of the class "The right process will produce the right results" to those principles. The *Toyota Way* then assigns two classes of principles, one for continuous improvement and one for process focus. On the other hand, the *Shingo Model* includes those two aspects in a single class called "Continuous (Process) Improvement". Concerning the relevance attributed to results, it's interesting to note that the *Shingo Model* attributes a class of principles to this aspect (Results) but the *Toyota Way* does not. It is also important to remember that in the *Lean philosophy* there is also a principle assigned to the identification and awareness of "value" for the customer.

In order to further explore the synergy between the models, we did the exercise of grouping, whenever possible, principles that seem equivalent (Table 2.3). As a reference for this grouping, the *Shingo Model* classification was used only because the description of each class is shorter. As can be observed in the referred table, the classes "Culture" and "Organization Alignment" are not represented by any principle of the *Lean philosophy*. Curiously, it is in the "Continuous Improvement" class that the focus on the process is also included, this being the class in which any of the models presents a greater number of principles.

Looking at Table 2.3 we recognize that most of the principles of both the Toyota Way and the Shingo Model belong to the less visible dimension of organizations, or in other words, to the "social" aspects of the socio-technical nature of organizations. Despite the possible differences in focus for each of the models and the use of different words and designations our position is that the lessons of each model should be understood as well as possible. We do not propose to use just one of the models as a guide on the journey to excellence but to explore the lessons of each of the principles that feature in each of the 3 models presented.

2.6 Guiding Principles of the KAIZEN Institute Model

In the same line of the principles from both *Shingo Model* and *Toyota Way* it might be acceptable to also refer the guiding principles of the *Kaizen Institute*. The Excellence in organizations model of this institute has influenced many important companies and organizations in Portugal, such as multinational industrial organizations, Portuguese logistics and distribution organizations, hospitals and many other important organizations of different sectors. The *Kaizen Institute* presents on its website (Kaizen Institute, 2020) the following guiding principles:

- Good processes result in good outcomes;
- Go see for yourself to understand the current situation;

Table 2.3 Equivalences between the principles of the Shingo, Toyota Way and Lean models.

Class	Shingo Model	Toyota Way	Lean
Culture	Respect every individual	**Principle 10**. Develop exceptional people and teams who follow your company's philosophy.	
		Principle 11. Respect your extended network of partners and suppliers by challenging them and helping them improve.	
		Principle 13. Make decisions slowly by consensus, thoroughly considering all options; implement decisions rapidly (*Nemawashi*).	
	Lead with Humility	**Principle 9**. Grow leaders who thoroughly understand the work, live the philosophy, and teach it to others.	
Continuous Improvement	Seek Perfection	**Principle 14**. Become a learning organization through relentless reflection (*Hansei*) and continuous improvement (*Kaizen*).	**Principle 5**. Perfection
	Embrace Scientific Thinking	**Principle 12**. Go and see for yourself to thoroughly understand the situation (Genchi Genbutsu).	
	Focus on Process	**Principle 6**. Standardized tasks are the foundation for Continuous Improvement and employee empowerment.	**Principle 2**. Value Stream
		Principle 7. Use visual controls so no problems are hidden.	
		Principle 8. Use only reliable, thoroughly tested technology that serves your people and process.	
	Assure Quality at the Source	**Principle 5**. Build a culture of stopping to fix problems, to get quality right the first time.	
	Flow & Pull Value	**Principle 2**. Create a continuous process flow to bring problems to the surface.	**Principle 3**. Flow
		Principle 3. Use 'pull' systems to avoid overproduction.	**Principle 4**. Pull
		Principle 4. Level out the workload (Heijunka).	
Enterprise Alignment	Think Systemically	**Principle 13**. Make decisions slowly by consensus, thoroughly considering all options; implement decisions rapidly (Nemawashi).	
	Create Consistency of Purpose	**Principle 1**. Base your management decisions on a long-term philosophy, even at the expense of short-term financial goals.	
Results	Create Value for the Costumer		**Principle 1**. Value

- Talk to the data, manage by facts;
- Take action to correct the causes of problems;
- Work as a team;
- Kaizen is everyone's responsibility.

In addition to these guiding principles, the creator of the *Kaizen Institute*, Masaaki Imai, in his book "*Gemba Kaizen: A Common-sense Approach to a Continuous Improvement Strategy*" (Imai, 1997) mentions the main *kaizen* systems, in which we identify two concepts covered by other typical principles of Excellence in organizations but not explicitly covered in the *Kaizen Institute*'s principles just listed. Those concepts are as follows:

- *Just-In-Time* (JIT) Production;
- Strategic Deployment.

The first aspect (JIT) is closely linked to the concepts of Flow and Pull while the second to the long-term vision and philosophy, properly aligned throughout all levels of the organization. Besides the explicit principles, both in this model (as well as in *Lean philosophy*), there are several concepts and practices which, although being aligned with many of the principles in the *Shingo Model* and *Toyota Way*, are not formally presented as principles. One example is continuous improvement, which, although not expressed in any principle of this model, is in fact its own very essence Kaizen (change for the better).

2.7 The Two Dimensions of the Models of Excellence in Organizations

Productive organizations are complex organisms which, to be successful and sustainable, require mastery of two major dimensions: the technical dimension and the human dimension. This type of recognition is not new and is often referred to as the socio-technical nature of organizations. Regarding the technical dimension we have the equipment, the technology, the organization and management of operations, the organization and management of logistics, the information systems, etc. In relation to the human dimension we have aspects like the culture of the organization, the behaviors, the values, the interpersonal relationships, the spirit, the motivation, the satisfaction at work, etc. To achieve Excellence in organizations it is necessary to make the most of these two worlds that are traditionally looked at and analyzed from different points of view. Organizations that excel in the technical dimension but ignore the human dimension are doomed to failure and we think that the same can be said about the opposite situation.

One of the big problems facing our economy is that experts in each of these dimensions are usually unaware of or undervalue the other dimension, resulting in an imbalance that clearly jeopardizes the success of many organizations. The humility of decision-makers and managers to recognize the value of the other party can be vital to the survival and success of their organizations. For organizations that manage to have teams of decision-makers with sufficient expertise to value both the technical and human sides, taking advantage of the natural synergies that are achieved with these two worlds, the achievement of success is greatly facilitated.

The way different scholars of the Toyota approaches, routines and practices, describe it, varies greatly in their emphasis on the technical aspects (physical/visible/tangible) and the human aspects (behavior/values/culture). If we analyze the principles of the 3 models highlighted in earlier, we can identify that some principles are more focused on the technical dimension (more visible) while others are more associated with the human or social dimension (less visible). Some of the principles could belong to both dimensions, since, depending on the point of view and assumptions, they can be physically manifested (can be observed) while in essence they can be intangible. Take examples like "Adopt scientific thinking" or "Pursuit of perfection". These and other essentially cultural and behavioral principles can also manifest themselves in a physical/visible medium, namely:

- In physical boards that may exist in the factory space with tools for structured problem solving;
- In exposed sentences that evidence this behavior;
- As part of some people's skills list and in training programs;
- Through people's behaviors and actions in workplaces;
- In the way people interact, greet each other and approach questions, and also how they show they are calm or stressed, happy or unhappy, interested or disengaged, etc.

In this context, we will use the terms visible side (or dimension) and less visible side (or dimension), without worrying too much about the accuracy of these expressions. The more visible side is the one that is easily observed in the space/place where the operations are performed and the less visible side concerns the aspects related to what is "inside" people and cannot be seen. But the truth is that this second side is not as easy to understand and interpret as the more visible aspects, such as the existence or not of intermediate stock, the existence or not of visual management, the level of tidiness of the premises, etc. The principles that in our view are more visible are highlighted in grey in Table 2.3, but other interpretations are of course can also be valid.

2.8 Tools Associated with Excellence in Organizations

Over the years it has become clear to my colleagues and me that most entrepreneurs and managers are more open to buying a solution or a tool than spending time understanding concepts, principles, models, philosophies and theories. It is perfectly understandable and also not expected or desirable for entrepreneurs and managers to be theoretical. On the other hand, in our teaching role it has always been easier to get students' attention for the implementation of tools than for the concepts and theory behind those tools. Well, the truth is that we don't have to be at either extreme and there will be cases where it makes sense to know more theory and there will be others where it makes more sense to buy the solution. The world of excellence in organizations seems to us to be that world where it is really very important to know the principles and concepts rather than simply buying a tool. There is a very curious phrase, attributed to Grady Booch, author of the book "*Software Engineering with Ada*" (Booch, Bryan, & Petersen, 1993), which says:

"A fool with a tool is still a fool"

In our area of knowledge this sentence applies quite well since most cases that are based on applying known tools (solutions) without their principle and concept being well understood result in failure. Anyway, the most popular tools and concepts applied in environments that seek excellence in organizations in line with the *Lean philosophy*, the *Shingo Model* or the *Toyota Way*, curiously are more easily related to only some of its principles. Table 2.4 presents some of these most popular tools as well as some concepts. There are cases where it is not easy to distinguish whether it is a concept or a tool. An example is for instance "*Standard Work*" which is both a concept and a tool. There are also concepts that do not translate directly into tools such as "*Going to the Gemba*", *Muda*, *Mura* and *Muri*.

As we can see in Table 2.4, a large part of these tools and concepts are associated with the principles "*Focus on Process*", "*Assure Quality at the Source*", "*Flow and Pull Value*", "*Seek Perfection*" and "*Embrace Scientific Thinking*". It is curious that most of the principles of the *Toyota Way* and the *Shingo Model* are more linked to the less visible dimension and this is also where there are fewer popularized tools.

Table 2.4 Most popular tools[7] and concepts and its relationship with some principles.

Principles	**Concepts / Tools**
Focus on Process / Assure Quality at the Source	5S*; *Andon**; Visual Management Boards; Work Standards; *Kamishibai**; *Poka-Yoke**; *Jidoka**; *Muri*; *Mura*; OEE*; Go too *Gemba*; VSM*
Flow & Pull Value	Supermarkets; *Mizusumashi**; Border of line*; *Kanban Systems**; SMED*; *Heijunka**; Cell Production; One-Piece-Flow; VSM*
Seek Perfection / Embrace Scientific Thinking	PDCA; *Toyota Kata*; A3*; Spaghetti Diagrams*
Create Value for the Customer	*Muda*; VSM*
Create Constancy of Purpose	*Hoshin Kanri*
Make decisions slowly by consensus, thoroughly considering all options; implement decisions rapidly	*Ringi Technique*
Think Systemically	VSM*; Process Mapping; PDCA

** A brief description of each of these concepts and/or tools is provided in the annex.*

Taking the principles of the *Lean Philosophy* as an example, four of them are dedicated to the most visible dimension (technical aspects) and only one to the less visible dimension (a mix of technical and human/cultural aspects). This is precisely the already mentioned principle of continuous improvement (or seek perfection), which is in fact the central theme of this book. In fact, the very attribution of the term *Lean* is more easily associated with physical aspects than with social or cultural aspects or respect for employees. The choice of the term *Lean*, in our interpretation, may have resulted from the researchers' perception of what they observed at the Toyota plants, namely the reduction of waste (fat) and of quantities of work in process (WIP), as well as the consequent increase in the flow of the articles throughout the manufacturing processes. As can be seen at the bottom of Figure 2.8, less WIP is metaphorically represented by a narrower pipeline, i.e. production can be said to "look leaner" (Lean).

As previously mentioned, it can be established that TPS includes a more technical side and a side more focused on human aspects and on the deliberate construction of an organizational culture more centered on the individual.

[7] A brief description of each of the tools in the table is provided in the annex.

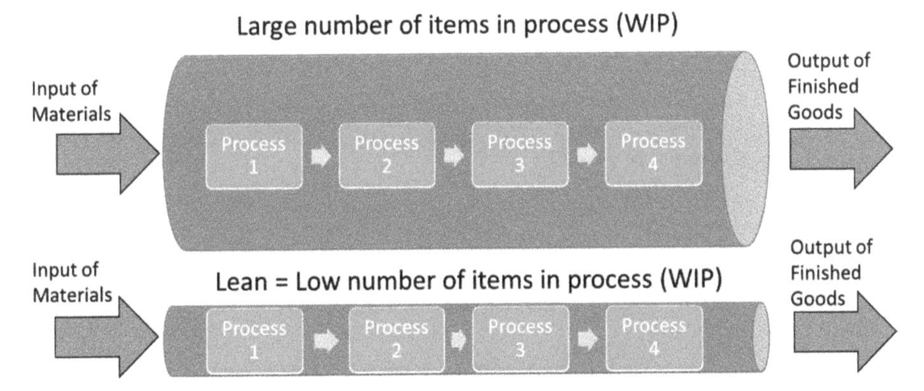

Figure 2.8 Leanness of lean philosophy.

The former includes techniques to reduce waste and increase customer-driven fluidity. These include tools such as *Kanban* and SMED techniques, *Poka-Yoke* devices and *Andon Systems*. On the more human side and organizational culture, there are aspects of strategy, vision, respect for people, behavior, leadership styles, teamwork and continuous improvement.

References

Beecroft, G. D., Duffy, G. L., & Moran, J. W. (2003). *The Executive Guide to Improvement and Change*. Qualilty Press.

Bhullar, A. S., Gan, C. W., Ang, A. J. L., Ma, B., Lim, R. Y. G., & Toh, M. H. (2014). Operational excellence frameworks-Case studies and applicability to SMEs in Singapore. In *IEEE International Conference on Industrial Engineering and Engineering Management* (Vol. 2015-January). https://doi.org/10.1109/IEEM.2014.7058722

Booch, G., Bryan, D., & Petersen, C. (1993). *Software Engineering With Ada*. Addison-Wesley.

Chakravorty, S. S., Atwater, J. B., & Herbert, J. I. (2008). The Shingo Prize for operational excellence: rewarding world-class practices. *International Journal of Business Excellence*, *1*(4). https://doi.org/10.1504/IJBEX.2008.018841

Coimbra, E. A. (2009). *Total flow management : achieving excellence with kaizen and lean supply chains*. [Howick, N.Z.]: Kaizen Institute.

Devine, F., & Bicheno, J. (2020). Creating Employee 'Pull' for Improvement: Rapid, Mass Engagement for Sustained Lean. In *Lecture Notes in Networks and Systems* (Vol. 122). https://doi.org/10.1007/978-3-030-41429-0_7

Fox, R. E. (1982). MRP, Kanban, and Opt - What's Best? In *Annual Conference Proceedings - American Production & Inventory Control Society*.

Goldratt, E. (1988). Computerized shop floor scheduling. *International Journal of Production Research*, *26*(3). https://doi.org/10.1080/00207548808947875

Goldratt, E., & Cox, J. (1984). *The Goal: A Process of Ongoing Improvement*. Great Barrington: North River Press.

Imai, M. (1986). *Kaizen: The Key To Japan's Competitive Success*. McGraw-Hill Education. Retrieved from https://books.google.pt/books?id=q0rCTQlvNMoC

Imai, M. (1997). *Gemba Kaizen: A Commonsense Approach to a Continuous Improvement Strategy. Library Journal*. https://doi.org/10.1080/10686967.2018.1404374

Johnson, C. (1997). Leveraging Knowledge for Operational Excellence. *Journal of Knowledge Management*, *1*(1). https://doi.org/10.1108/EUM0000000004579

Jones, D., & Womack, J. (2002). *Seeing the Whole: Mapping the Extended Value Stream. Lean Enterprise Institute, Brookline*. Cambridge, MA, USA: Lean Enterprises Inst Inc.

Kaizen Institute. (2020). Kaizen Institute. Retrieved from https://www.kaizen.com/

Kamann, D. J. F. (2007). Organizational design in public procurement: A stakeholder approach. *Journal of Purchasing and Supply Management*, *13*(2). https://doi.org/10.1016/j.pursup.2007.05.002

Kidd, P. (1994). *Agile Manufacturing: Forging New Frontiers*. Addison-Wesley.

Krafcik, J. F. (1988). Triumph of the lean production system. *Sloan Management Review*, *30*(1), 41–52. https://doi.org/10.1108/01443570911005992

Liker, J. (2004). *Toyota Way: 14 Management Principles from the World's Greatest Manufacturer*. McGraw-Hill Education.

Liker, J., & Franz, J. (2011). The Toyota Way to Continuous Improvement : Linking Strategy and Operational Excellence to Achieve Superior Performance. McGraw-Hill Publishing.

Miller, R. D. (2018). *Hearing the voice of the Shingo principles : creating sustainable cultures of enterprise excellence*. Routledge.

Nagel, R. N., & Dove, R. (1991). *21st Century Manufacturing Enterprise Strategy, An Industry-Led View*. (L. U. Iacocca Institute, Ed.). Bethlehem, PA: Diane Pub Co.

Ohno, T. (1988). *Toyota production system: beyond large-scale production*. (C. Press, Ed.) (3ª Edição). New York: Productivity, Inc.

Plenert, G. J. (2017). *Discover excellence : an overview of the Shingo model and its guiding principles*. New York: CRC Press.

Productivity Press Development Team. (1998). *Just-in-Time for Operators (The Shopfloor Series)*. Productivity Press.

Pyzdek, T. (2003). *The Six Sigma Handbook Revised and Expanded*. McGraw-Hill. https://doi.org/10.1036/0071415963

Rai, A., Patnayakuni, R., & Seth, N. (2006). Firm performance impacts of digitally enabled supply chain integration capabilities. *MIS Quarterly: Management Information Systems*, *30*(2). https://doi.org/10.2307/25148729

Rother, M., & Shook, J. (1999). *Learning to see: Value stream mapping to add value and eliminate muda. The Lean Enterprise Institute*. https://doi.org/10.1109/6.490058

Sagi, S. (2015). "Ringi System" The Decision Making Process in Japanese Management Systems: An Overview. *International Journal of Management and Humanities*.

Senge, P. M. (2006). *The fifth discipline : the art and practice of the learning organization*. Doubleday.

Shannon, P. (1997). The value-added ratio. *Quality Progress*, *30*(3), 94–97.

Shingo Institute. (2020). The Shingo Model. Retrieved June 8, 2020, from https://shingo.org/shingo-model/

Shingo Institute. (2106). Shingo Institute: Our history. Retrieved September 6, 2021, from https://www.cob.calpoly.edu/centralcoastlean/wp-content/uploads/sites/6/2017/07/Shingo-California-presentation-Feb-2016-edited.pdf

Shingo, S. (1985). *A Revolution in Manufacturing: The SMED System*. Oregon: Productivity Press.

Shingo, S. (1989). *A study of the Toyota production system from an industrial engineering viewpoint*. New York: CRC Press.

Sony, M. (2019). Implementing sustainable operational excellence in organizations: an integrative viewpoint. *Production & Manufacturing Research*, *7*(1), 67–87. https://doi.org/10.1080/21693277.2019.1581674

Sugimori, Y., Kusunoki, K., Cho, F., & Uchikawa, S. (1977). Toyota Production System and Kanban System: Materialization of Just-in-Time and Respect for Human Systems. *International Journal of Production Research*, *15*, 553–564. https://doi.org/10.1080/00207547708943149

Thürer, M., Tomašević, I., Stevenson, M., Fredendall, L. D., & Protzman, C. W. (2018). On the meaning and use of excellence in the operations literature: a systematic review. *Total Quality Management and Business Excellence*. https://doi.org/10.1080/14783363.2018.1434770

Toyota_Europe. (2019). The Toyota Way: our values and way of working. Retrieved January 20, 2020, from https://www.toyota-europe.com/world-of-toyota/this-is-toyota/the-toyota-way

Toyota_Europe. (2020). Toyota Global Vision and guiding principles. Retrieved September 29, 2020, from https://www.toyota-europe.com/world-of-toyota/this-is-toyota/toyota-global-vision

Toyota_Motor_Corporation. (2009). Sustainability Report 2009: Social Aspects. Retrieved September 15, 2021, from https://web.archive.org/web/20130731092215/http://www.toyota-global.com/sustainability/sustainability_report/pdf_file_download/09/pdf/sr09_p54_p59.pdf

Toyota_Motor_Corporation. (2012). 75 Years of Toyota. Retrieved September 15, 2021, from https://www.toyota-global.com/company/history_of_toyota/75years/index.html

Womack, J., & Jones, D. (1996). *Lean thinking: Banish Waste and Create Wealth in Your Corporation*. New York: Fee Press.

Womack, J., Jones, D., & Roos, D. (1990). *The machine that changed the world*. New York: Free Press.

3

The Less Visible Side

The less visible side of the organization and management of companies is on the human side, in behavior and culture. This side, or dimension, is present in all types of companies and organizations, but, in most cases, it is not taken care of at all and leaders do not pay it due attention. For an organization to have sustainable success, it is necessary that this less visible side is planned, structured and materialized on a daily basis, in all areas of the organization. This chapter will emphasize what organizations can contribute to satisfying the needs of people as human beings, and also what human beings, in addition to performing operations, can contribute to organizations with creativity, innovation, commitment and team spirit.

The invisible side of organizations that was referred to, for instance by Mike Rother (Rother, 2010) by saying that the part of Toyota that is most difficult to understand and copy is the one linked to people's culture and behaviors. This human side has been addressed in many publications over the years, but the first reference was most likely most likely in the first article in English language published about TPS (Sugimori et al., 1977). In that article, the authors distinguish a more visible side as being the concept of reducing costs by reducing waste (non-value-added activities), and a less visible side which they called "treating employees as human beings and with consideration". A further reference to these two sides of organizations is also made by Daniel Fleming from the Shingo Institute in a very interesting presentation that is recorded as a video on Youtube (Fleming, 2018). In this presentation of the Shingo Model, Fleming brilliantly establishes the differences between the more physical side of tools and results and the side of principles and behaviors. On the physical side he refers to KPI (*Key Performance Indicators*), while on the principles and behaviors side he refers to KBI (*Key Behavior*

Indicators). Human aspects are, without a doubt, a vital dimension for the success of organizations. Today it is accepted by many that in the long run what makes the difference in the success of some companies and organizations is the fact that they focus on the full understanding of principles and behaviors. The existence of that side, or that less visible dimension of Toyota's approach, has been recognized and noticed by many people, but it has not been easy at all for other companies and organizations around the world to achieve the same kind of effectiveness.

The less visible side of Toyota's approach cannot be observed effectively in just a few visits. Observing only the physical production spaces, it is not possible to identify aspects such as culture and the existence or not of: team spirit, commitment, respect for the individual, use of human capital in its fullness, self-fulfillment and self-esteem among employees.

3.1 The Employee as a Human Being

The principle of Continuous Improvement that is present in the three models of excellence in organizations presented in the previous chapter is understood in this book as the main guarantee of the future survival of organizations. As we will see throughout this work, Continuous Improvement is also intimately linked to the full use of workers' capabilities and, curiously, also linked to the need to "treat workers as human beings and with consideration". This last aspect is one of the key concepts mentioned in the first publication in English language about the Toyota Production System (Sugimori et al., 1977) and which is later also referred to as a pillar of the Toyota Way, with the name Respect for People. In the same vein is the Shingo Model's "*Respect Every Individual*" principle. Although in the *Toyota Way* the pillars "*Respect for People*" and "Continuous Improvement" are separate, we understand that Continuous Improvement is already in itself a form of respect for the individual. However, it is legitimate to ask how Continuous Improvement can be so closely linked to respect for people and to the use of human capital to its fullest?

On the subject of treating staff with respect, during years we asked many owners and top managers of many organizations if they treated their staff as human beings and with consideration. The answers were given quickly and without any hesitation with an obvious "yes". Many even gave several practical examples to prove it, and in many cases we were convinced. To our surprise, when we questioned employees led by these same owners and top managers, we learned that many did not feel like they were treated with respect and consideration. This discrepancy in point of view created in us a great need to better understand this issue. The same reality was perceived very differently depending on whether it was from the point of view of the owner of the company (sender) or the point of view of the employee (receiver). The question that is perhaps most pertinent to ask is: what conditions does an employee need to have in order to feel respected and treated as a human being in his or her fullness?

To try to answer this question we should do the exercise of trying to identify the requirements for a person to feel respected as a fully human being. In other words, what might our needs be as humans to feel complete? If we identify these needs, then perhaps we can understand what organizations can do to get the most out of human capital (or treat employees as human beings and with consideration).

A possible answer can be found in the so-called hierarchy of needs proposed by the American psychologist Abraham Harold Maslow (Maslow, 1943), which can be represented by the pyramid illustrated in Figure 3.1 (here slightly adapted from the original). This pyramid maps human needs in a hierarchical way. On the right side of the figure some characteristics associated with the work context and which are aligned with Maslow's proposal are presented. According to this hierarchy of needs, the motivation to satisfy the needs of a higher level arises when the lower level is reasonably satisfied. According to this model, the major objective is to reach the highest level (top

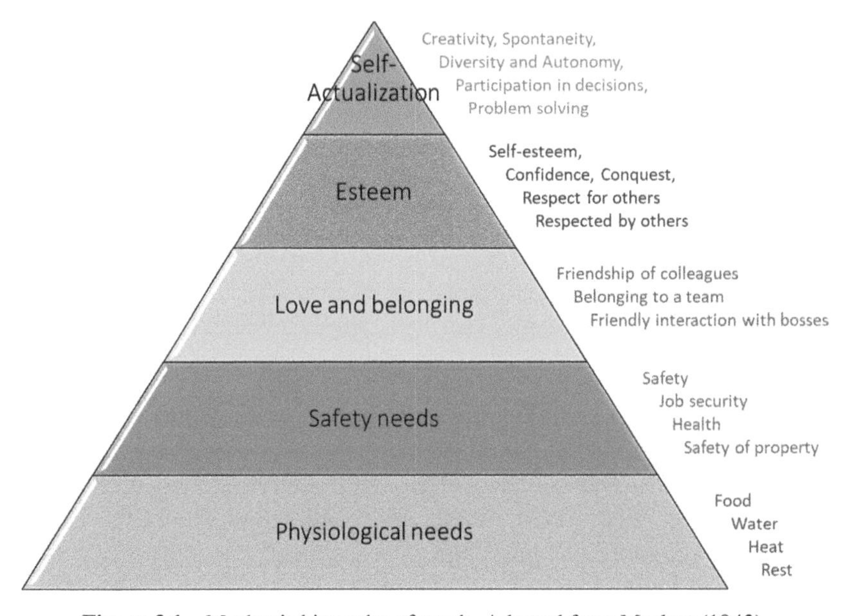

Figure 3.1 Maslow's hierarchy of needs. Adapted from Maslow (1943).

of the pyramid), that is, to achieve personal fulfilment. Another interesting aspect that emerges from Maslow's proposal is that people always want more and what they want depends on what they already have.

The most basic needs, at the base of the pyramid, are the physiological needs and include homeostasis (automatisms of the body to maintain its normal functioning such as regulating the amount of water and minerals in the body, body temperature and blood sugar levels), satisfaction of hunger and thirst, sleep, sensory pleasures, activity and possibly sexual desire. In the context of work, a large part of these needs should be guaranteed and framed in ergonomically related working conditions, namely the thermal environment and air quality, breaks and access to water and food.

If these basic needs are satisfied, then the human being becomes focused on security needs. These include safety and security from the threat of attack, protection from danger or deprivation, the need for predictability, having a job, having health and a sense of ownership. In the work context, some of these issues are usually difficult to guarantee, in particular the issue of job security and predictability. In a large part of the organizations, there is no guarantee of job security, but there are other aspects of security that organizations can guarantee, for example personal safety and health. In fact, there is a wide spectrum of actions aimed at ensuring maximum safety for employees and eliminating the risk of accidents. The area of industrial safety is totally focused on this type of human need and includes, for example, the implementation of rules, procedures and equipment that guarantee the safety of employees. In addition to accident prevention, this also includes preventing the risk of developing musculoskeletal disorders, as well as other risks associated with the development of other occupational illnesses. Toyota dedicates part of this type of risk exposure to a concept called *Muri*, as mentioned by Jeffrey Liker (Liker, 2004) and already presented in chapter 1. The word *Muri*, which is often translated as overload, and is also used in this context to refer to repetitive, very demanding or risky tasks, and the excessive stress/strain required to carry out a task or operation. To expose a person to *'Muri'* is to push that person beyond their natural limits; this can occur when the manager tells employees to work harder (Stewart, 2011).

After securing the first two levels of needs, human beings begin to pay attention to the needs inherent in their relationship with others. This level includes the need to give and receive love (often referred to as social needs), the feeling of belonging, participation in social activities and the need for friendships and a family. A good part of these needs can and should also be satisfied in the work environment, namely the aspects of belonging, social activities and friendship. The feeling of belonging to a team, a department or a organization is extremely important, it should be promoted within organizations and effective Continuous Improvement environments are one way to achieve this. Social activities should also be promoted, for the well-being of people and organizations. As an example, take a look at the many technology organizations that currently make good use of it as a way to develop and enhance team spirit (some of them being often classified as "the best organization to work for").

At the next level (Figure 3.1), esteem needs (also sometimes referred to as ego needs) arise. This class of needs includes self-respect and the esteem of others. Self-respect involves the desire for confidence, strength, independence, freedom and achievement. Esteem of others involves reputation or prestige, status, recognition, attention and appreciation. In the work context, this level of needs is well aligned with Continuous Improvement, since it promotes the existence of teams with a high degree of autonomy in terms of decision-making, local management and problem solving. In fact, and Continuous Improvement involves this very thing, for these teams it is very important this empowerment, the participation in management tasks and in the creation of goals and challenges, and, not least, the existence of a system of recognition and appreciation of their achievements.

Finally, at the top of Maslow's pyramid are the needs for self-actualization. Involved here is the harnessing (development and realization) of the full potential that each person has. Maslow sees this as "What a human being can be, should be" or "a human being should become all that he is capable of becoming". Self-actualization needs can take many forms, which vary widely depending on the individual. In the work context, organizations should create an environment that brings together the right conditions for each of the employees to achieve personal satisfaction that is possible in work contexts.

In conclusion, if we use this Maslow model, treating employees as human beings and with consideration can be fully achieved if all five levels of needs mentioned and presented in Figure 3.1 are satisfied. It will certainly not be easy, but if an organization can at least pursue these goals, then it will already be on the right track. Organizations, in order to be successful, must create conditions for the worker to feel safe and socially integrated, to have self-esteem and respect from others, to be exposed to challenges and to feel independent and growing professionally/personally. Continuous Improvement principle help to achieve these major objectives.

Employees should include in their work routines, not only the specific operations of their jobs, but also their contributions to management and improvement. Those are precisely two example where the concept of treating the employee as a human being and with consideration can "fit in". It is with teamwork and with the right framework for Continuous Improvement that the necessary conditions can be created for all employees to feel as human beings in their fullness.

In many industrial realities, the work itself (the routine tasks) carried out by direct employees may not be interesting enough to generate stimulation, motivation, job satisfaction, self-esteem, a sense of recognition, etc. However, we also know that in the overwhelming majority of cases it will not be possible to significantly change the routine tasks in order to make them interesting enough to make use of the human potential in its fullness. So, in these cases, the challenge will be: how to create a suitable framework for human needs to be met?

3.2 Motivation and Job Satisfaction

The concept of work motivation and the concept of job satisfaction are often confused, but it is very important to understand what makes them different. If you want to be a more effective leader, you should try to understand these differences very well, in order to get the best out of your employees, while serving them better and giving them a better quality of life. Job satisfaction is related to the conditions that are given to the employee. An employee can be very satisfied because the organization provides him with a good salary and a good schedule, job security, health insurance and a nursery school for his children. At the same time, that same employee may be doing extremely monotonous and tedious tasks, which completely bore and demotivate him/her. The opposite can also occur; it is possible for a person to have a highly challenging and motivating job, always overcoming obstacles and solving problems, developing enormously as a professional and as a person, but nevertheless having a bad salary and/or inadequate conditions for the work they do.

Some leaders/managers believe that they increase the motivation of their employees by improving the conditions just mentioned, but in fact they can only improve their employees' satisfaction. It is obviously desirable to improve job satisfaction, but if only this dimension is worked on, ignoring the motivation dimension, the organization will have great difficulty in improving its performance and surviving in the medium/long term. There are hundreds of important organizations that, although providing job satisfaction, failed precisely in this dimension of employee motivation and ended up disappearing from the market.

There are several ways to promote employee motivation, but the one we will use in this book is closely related to Continuous Improvement. The idea is to involve employees in the management of the areas where they work and to make them feel like integral members of an organization, playing an important role for the whole. By involving and engaging the employees in an integrated system of Continuous Improvement we will be creating the adequate environment so that they can feel as fully human beings. With a Continuous Improvement system we create conditions for the workers to be able to:

1. Face challenges;
2. Belong to a social group - a team;
3. Obtain recognition for their achievements;
4. Improve their self-esteem;
5. Respect and be respected by others;
6. ...

As we will see later, teamwork is a key factor in Continuous Improvement work as well as in the motivation of workers and managers. The positive effects of teamwork are reported in several studies published over the years, namely in terms of performance improvement as proven by (Delarue, Van Hootegem, Procter, & Burridge, 2008) and (Benders, Huijgen, & Pekruhl, 2001). In addition to this type of improvement, there are other quite interesting results. For example, self-managed teams have obtained significant improvements in job satisfaction when compared to traditional work groups or departments (Cohen & Ledford, 1994). Good results in reducing absenteeism were also noted by (Benders et al., 2001). Rosemary Batt (Batt, 2004) showed in her article that self-managed teams were associated with significantly higher levels of insight, job security, and job satisfaction, and were more effective in improving objective performance measures.

Continuous Improvement and teamwork are very present in the Toyota Production Model. If we look at the house of the Toyota Production System in Figure 3 (chapter 1), it's easy to observe that, as central elements (between the two pillars of the house), the expressions "Continuous Improvement", "people and teamwork", "versatility" and "*Ringi* decision-making" clearly appear. The latter represents a very important dimension in Continuous Improvement, the decision making based on consensus. When the *Ringi* technique[1] is applied, ideas and plans are discussed, developed and refined in informal meetings between employees, not aimed at majority-based decisions but at consensus-centered decisions (Sagi, 2015).

[1] Ringi technique is briefly described in the annex.

One of the 14 *Toyota Way* principles says: "Make decisions slowly by consensus, considering all options in detail; implement them quickly". Decision-making by consensus takes more time than decision making by just one leader, but its implementation is much more effective. This is a very important learning for the success of Continuous Improvement.

Continuous Improvement, or the pursuit of perfection, has an important place in the *Lean Thinking* model proposed by (J. P. Womack & Jones, 1996) in their "*Lean Thinking*" book, since it appears as one of the 5 principles of this model. The importance of Continuous Improvement as well as teamwork had already been mentioned in the Lean Production approach presented in the famous book "*The machine that changed the world*" (J. Womack et al., 1990). Despite this recognition/warning, the truth is that the detail with which these themes were explored in these books was not enough to effectively help organizations/managers to effectively implement continuous improvement systems in their realities.

Teamwork, although it continues to be looked upon with some levity and disinterest by many organizations, has already been a concept formally applied for many decades. We use the term "levity" because many organizations, although mentioning through speech and written messages in walls the importance of teamwork, do not actually have any type of organization and management structured in teams. Many organizations even promote impressive training courses, such as *Teambuilding*, but, in reality, they systematically refuse the practical implementation of teamwork and instead preserve a hierarchical structure of people with objectives and direct instructions for each employee.

It is important to note that, obviously, it is not enough to form teams and give them autonomy to suddenly make improvements happening in the organization. It is very important to know how to build the necessary environment to allow real teamwork to exist effectively and bring benefits to everyone and the organization. This subject will be covered later in this book.

The so-called Quality Circles are probably pioneers in the effective use of formal work in operational teams in industry. They apparently originated in the 1940s of the last century (Nonaka, 1993), although the term was first used in 1962 in Japan. Quality Circle can be defined as being "a group of workers, usually 8 to 12, working in the same area or doing similar work, formed on a voluntary basis to solve problems related to their work" (Kulkarni et al., 2018). These groups, usually led by a supervisor or manager, present their solutions to top management and, where possible, the workers themselves implement the solutions to improve the organization's performance and everyone's motivation. These groups of workers solve problems not only inside but also outside their working hours (Lawler_III & Mohrman, 1985). This topic of workers being available to think and solve problems outside of working hours is a very interesting one. The reader will realize that this does not always happen in organizations, and when it does, it can be an indicator of how motivated employees are to work and their level of commitment to the organization's goals.

3.3 Resistance to Change

There is yet another aspect, always present in Continuous Improvement initiatives and mentioned by many, that requires some attention: resistance to change. It is undeniable that an organization that intends to implement Continuous Improvement has to be prepared for that phenomenon. Resistance to change should be accepted naturally; otherwise it could lead to frustrating situations, and, in that sense, it might be important to try to understand a little better this kind of behavior.

If it is true that resistance to change is often evoked as being present in human nature, it is no less true that the will to change is also present. In fact, these two truths are only seemingly contradictory. In a sense, all people are willing to change. The reader will probably agree that occasionally some people want to change cars, others want to change jobs, others change cities, etc. In reality, we are willing to make changes, as long as they result from our own will. Most of us are willing to go through quite radical changes such as being willing to get married or decide to have children even though we know that these changes completely alter our lives. Having children causes drastic changes in our lives, such as changes in routines, sleepless nights, constant worrying, constant attention, fear of the unknown, visits to the doctor, and much more. Even with this willingness to change, we are not always willing to change when others, other than ourselves, want us to change something in our routines, behavior, or actions. According to a very interesting work on this subject, the book "*Switch: How to Change Things When Change Is Hard*" (Heath & Heath, 2010), for people to change, three conditions must be met: (i) clarity on change; (ii) motivation for change and (iii) clarity on the steps necessary for change. Regarding the first condition, sometimes it is not very clear to people what needs to change, and this, of course, is undesirable. When you want a team to make a change, you have to make it very clear what the change is going to be. As far as the second condition is concerned, you need the people who are going to be involved in the change to have the required motivation to make it. If they do not identify something that is attractive or worthwhile, then those people will not want to change. This is not necessarily a selfish position of those involved. In fact, people may be motivated by a change that brings benefits to others (and not to themselves) or that is aligned with a belief, a principle, or simply because it is a cause for which they are willing to make the change. Lastly, it is also necessary that people understand very clearly the steps necessary for a change to take place. Even if people know the reason for the change and have every motivation to make the change, they may not be able to make the change if they do not know exactly what needs to be done.

3.4 The Role of the Organization's Leader

The way the company/organization integrates the different entities and the different roles of Continuous Improvement (CI) in its organizational structure shows how it understands and wants it. In organizations that formally include CI in their organizational structure it is possible to find different approaches. One type of approach that is quite common, especially for organizations that are just starting their initiatives, is based on assigning the responsibilities for CI to one person or a group of people in the organization. Often, organizations dub this type of initiative as "implementing Lean" or "implementing *Kaizen*" although it does not always work out positively. Although there are no solid studies to support a number and a reason, it is estimated that most attempts to implement Lean fail. The reason for this failure is the same reason why many people fail when adopting a diet to lose weight: lack of attitude change. It is the same reason why, for example, many people try techniques and therapies to quit smoking but fail. Just as it is necessary to change one's eating habits and exercise routines in order to lose weight, so it is also necessary in organizations to change the way tasks are performed, the various routines and habits, thus changing the way of thinking and, consequently, the culture of the organization. The reader will probably agree that changing organization's routines is no easy task and requires unwavering motivation/determination/will on the part of the organization's top management. The degree of commitment and involvement of top management, the business owner, the leader or "number 1" as they refer to (Bastos & Sharman, 2018), in the process of changing to a Lean culture is perhaps the most important factor for success. In the same vein, the creator of the Kaizen Model, Masaaki Imai, often mentions in his lectures that the 3 critical success factors for a transformation to Kaizen (or Lean) are: "Top management determination, top management determination, and top management determination" (Graban, 2012). This means, very simply, that the top leader of an organization must commit to Continuous Improvement in the same way that a person must commit to exercise and diet changes in order to gain fitness. It is not enough for a person to pay gym membership fees and assign a personal trainer the task of getting that person in shape. One must go to the gym at the frequency set by the personal trainer, one must perform the exercises according to instructions, and one will probably need to adopt new sleeping and eating routines in order to get in shape. Just as it is necessary to change routines and habits to achieve improved fitness, so it is necessary for the entire organization to change routines and habits to achieve Continuous Improvement and operational or organizational excellence. The organization's leader is by his/her role the person who can have no doubts about the path to take and will have to be committed to be part of the necessary transformation, also changing his/her habits and routines, so that the transformation in the organization in terms of CI becomes a reality. Trying to implement CI without the organization's leader being clearly committed, not only with words, but specially with actions and attitude, is a very demanding task and with a very high probability of failure. That said, it is advisable that anyone who is motivated to start including CI routines in their organization should start by making the leader understand and really believe in the expected benefits and what it implies in terms of changes in the organization.

3.5 Learning Organizations

There is an aspect that has probably gone unnoticed, but which is very interesting and curious. It is principle 14 of the *Toyota Way*, described as follows: "Become a learning organization through relentless reflection (*Hansei*) and Continuous Improvement (*Kaizen*)". The word "*Hansei*" suggests self-reflection, and in Japanese culture it means to admit one's mistake, learn from it, and improve. It is common practice in Japan that, for example, a politician who commits a crime of corruption admits his mistake in public and steps aside for a few years. It is however acceptable for that person to return to politics later because the community believes that they will have learned their lesson.

Learning in organizations is an aspect that is becoming increasingly important and that has given rise to the famous term "*Learning Organizations*".

In this book the term "*Learning Organizations*" will be adopted because it is already used frequently. A possible definition is: a learning organization is an organization "with the ability to learn faster than its competitors" (Senge, 2006). As far back as 40 years ago, Reginald Revans already argued, in his book entitled "*Action Learning: New Techniques for Management*" (Revans, 1980), that the rate of learning should be greater than (or at least equal to) the rate of change within the organization. Still on this theme of learning, but much more recently, Powel and Reke argue, in an article with the suggestive title "*No Lean Without Learning: Rethinking Lean Production as a Learning System*" (Powell & Reke, 2019), that the focus should be on learning and not on pure process improvement. The focus should be on developing personal skills (both technical and creative) at all levels of the organization (i.e., from employees to top managers).

The underlying idea is that learning organizations assume that continuous learning is part of their culture. All the people in these learning organizations are continuously learning and becoming better people. The Shingo Model assumes that in a highly competitive organization, the full potential of each individual must be achieved. In this model people are considered a very special asset, in that they are seen as being the only organizational asset with infinite capacity. This way of thinking is very relevant. While all other resources are limited by their nature, people have no limit to their growth, capacity for innovation, creativity, etc. The challenges inherent in competing in global markets are so great that success is only achieved when all individuals at all levels of the organization are able to innovate and improve continuously.

Continuous Improvement is very probably the most important dimension of the Toyota approach that has been captured by the *Lean Thinking* Model in one of its principles, and it depends very much on cultural and human aspects that will have to be developed in organizations. Continuous Improvement, very much linked to the principle "*Seek Perfection*", is perfectly aligned with the concept of learning organizations. Continuously improving leads to continuously learning, and vice versa. Moreover, it can also be argued that an effective way to achieve motivation and job satisfaction of people will be through the effective implementation of a Continuous Improvement system. Teamwork, autonomy and responsibility are very powerful ingredients to achieve work motivation, Continuous Improvement and to guarantee the future of an organization. It is very interesting how Jeffrey Liker condenses in a single paragraph a very interesting set of statements about the creation of a Continuous Improvement culture.

"Of course, you cannot pull a ready-made culture out of a wizard's hat. Building a culture takes years of applying a consistent approach with consistent principles. It includes the foundational elements of Maslow. People must have a degree of security and feel they belong to a team. You must design jobs to be challenging. People need some autonomy to feel they have control over the job. Moreover, there seems to be nothing as motivating as challenging targets, constant measurement and feedback on progress, and an occasional reward thrown in. The rewards can be symbolic and not all that costly. In the end, building exceptional people and teams derives from having in place some form of a 'respect for humanity system'." (Liker, 2004)

An interesting example is the following: The number of suggestions proposed by employees in specific plant, of a large French group in the automotive sector with some plants in Portugal, suffered a drastic decrease in a certain period some years ago. After trying to find out the causes of this drastic reduction, top managers found out that there were rumors circulating that there would be layoffs due to reduced demand. The lack of security in keeping the job generated a huge demotivation in most employees and with it a lower willingness to contribute to continuous improvement. After some meetings with employees showing that there was nothing real about the rumors, the number of suggestions gradually increased to normal levels.

Going back to the Jeffrey Liker, he also mentions that people need some autonomy to feel that they have control over their work. Autonomy is again mentioned as being a very important ingredient for motivation and job satisfaction. In addition, there seems to be nothing quite as motivating as setting challenging goals, constant monitoring and feedback on progress, and an occasional reward. When challenging goals are set it is necessary to keep in mind that the challenge should be neither too easy nor too difficult to achieve. Mike Rother, in his book "*Toyota Kata*" (Rother, 2010), discusses how these challenges should be assigned/negotiated/assumed. Each challenge must be negotiated by the team responsible for achieving the goal and the manager or facilitator (Coach) so that it is neither too easy (green area in Figure 3.2 (a) nor too ambitious (red area in the same figure). If it is too easy, the team does not find enough motivation and, obviously, does not learn. If the challenge

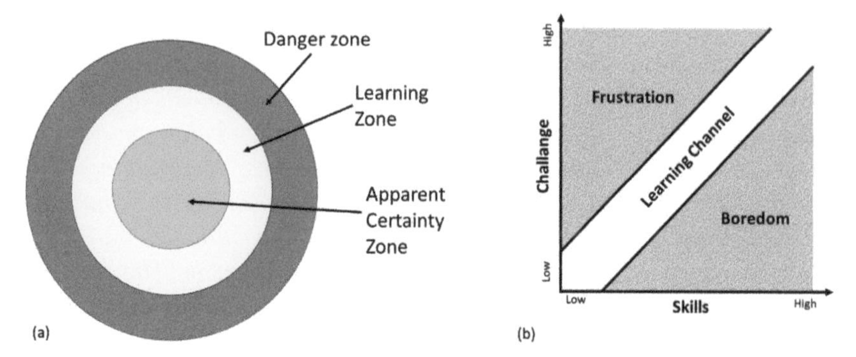

Figure 3.2 Learning zone (Rother, 2010); (b) Csikszentmihalyi model (Csikszentmihalyi, 1990). Adapted from Csikszentmihalyi (1990) and Rother (2010).

is too demanding, people also lose motivation because they feel they do not have enough ability, and because they do not try to solve it, they do not learn either. When the challenges are evenly defined in terms of ease/difficulty, people are motivated, and the experience resulting from solving those challenges generates learning.

In the same line of thought, there is another model that resorts to a slightly different visualization, as can be seen in Figure 3.2 (b). Besides the degree of difficulty of the challenge, this model considers another dimension - the level of skill/knowledge/experience of the person -, and explores the relationship between these two dimensions. As can be observed, feelings of frustration and boredom are strongly associated with this relationship. In fact, the same challenge, when proposed to people with different skill levels, may generate boredom in some, frustration in others or motivation (generating learning) in still others. There is a kind of learning channel, called the "Flow channel" in Csikszntmihalyi's original model (Csikszentmihalyi, 1990), which is the zone where a person exposed to a given level of difficulty (of a challenge) is motivated and learns (provided that his or her level of ability is appropriate to that challenge).

3.6 Engagement of Everyone

The level of engagement felt by an organization's employees is a vital factor in that organization's success and long-term sustainability. It is however important to note that engagement alone is not enough; the organization must also have a viable project, an inspiring vision and appropriate values. However, it is almost certain that a lack of employee commitment makes success impossible, even if all the other aspects just mentioned are ensured.

> Engagement should not be confused with involvement. An interesting way to clarify the difference is to use the metaphor of the typical English breakfast with bacon and eggs. In this metaphor, the pig is clearly engaged while the chicken is only involved.

We often talk about the importance of employee involvement, but in fact, organizations should try to go a bit further; they should look not only for ways to involve employees, but also to make them feel engaged with the organization. When everyone in a organization is engaged, then it is expected that "good things" will happen. One question may arise: how can the engagement be assessed.

One of the ways of assessing people's engagement with their organizations is the rather famous questionnaire created by the Gallup Organization (www.gallup.com). This questionnaire (*Gallup Enquiry*) contains only 12 questions that apparently assess very effectively the degree of a person's engagement with its company or organization. The CEO of the Gallup Organization, James Clifton, states, "The success of your organization does not depend on your understanding of economics, organizational development or marketing. It depends, simply, on your understanding of psychology: how each employee connects with your organization and how each individual connects with your customers" (Hines et al., 2011).

The questions in this questionnaire can be answered on a Likert scale and are as follows:

1. *Do you know what is expected of you at work?*
2. *Do you have the materials and equipment to do your work right?*
3. *At work, do you have the opportunity to do what you do best every day?*
4. *In the last seven days, have you received recognition or praise for doing good work?*
5. *Does your supervisor, or someone at work, seem to care about you as a person?*
6. *Is there someone at work who encourages your development?*
7. *At work, do your opinions seem to count?*
8. *Does the mission/purpose of your company make you feel your job is important?*

9. *Are your associates (fellow employees) committed to doing quality work?*
10. *Do you have a best friend at work?*
11. *In the last six months, has someone at work talked to you about your progress?*
12. *In the last year, have you had opportunities to learn and grow?*

According to the Gallup organization, there are three types of employee: the engaged, the disengaged and the actively disengaged. The description of each of these three types is as follows:

- **Engaged**: the employee works with passion and feels a deep connection to the organization. The employee manifests by his/her behavior an intrinsic interest in the work and the organization;
- **Not Engaged**: the employee works just to do the time without giving energy and passion to the work. This employee can even be satisfied with his/her job, but does not show any intrinsic interest in either the work or the organization;
- **Actively Disengaged**: the employee is not only unhappy with the job, but constantly seeks to demonstrate this unhappiness. This employee undermines what his engaged colleagues, with work and with the organization, can achieve.

The way in which the Gallup organization classifies an employee according to the answers to the survey is not public knowledge and therefore it will not be possible to describe here how the classification of "engaged", "not engaged" or "actively disengaged" is assigned. Nevertheless, and assuming that the assignment of each of the mentioned connotations is reliable, according to a study carried out by the Gallup organization between 2011 and 2012 (Crabtree, 2013) only 13% of workers worldwide are engaged with their work and with their organizations. In Figure 3.3, the reader can observe the distribution of the degree of engagement of workers in the various regions of the world.

Still on the same study, and with regard to some European countries, we have the values presented in Figure 3.4. Denmark, in relation to its closest partners, appears well positioned in the percentage of employees engaged with work and with their organizations. As a curiosity, according to this study, Croatia has a very low level of committed employees (only 3%). We do not know what this actually means, as we do not know the statistical certainty of

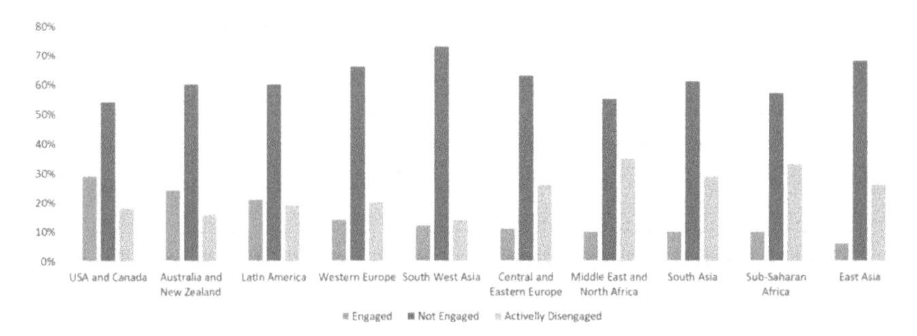

Figure 3.3 Engagement levels in different regions in the world. Adapted from Crabtree (2013).

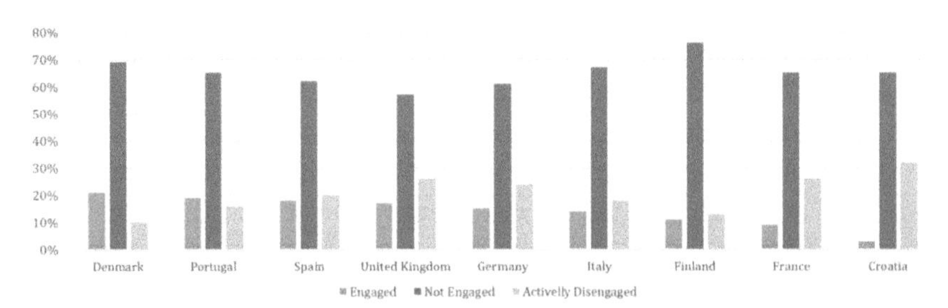

Figure 3.4 Engagement levels in some European countries. Adapted from Crabtree (2013).

the study and furthermore there may be cultural issues or the way the questions are interpreted by people that bias these results.

> *"When people are financially invested, they want a return. When people are emotionally invested, they want to contribute."*
>
> (Simon Sinek in (McLeod, 2013), p. 322)

This issue of employee engagement in organizations has been effectively addressed by Frank Devine. He has created a methodology called Rapid Mass Engagement that aims to rapidly increase the level of engagement within any organization (Devine & Bicheno, 2020). According to him worker´s engagement is very connected to the level of their involvement in creating the corporate culture. Normally, organizations try to sell their corporate culture to their employees; with Rapid Mass Engagement process employees create and own their own culture.

3.7 Organization's Vision

The vision of a organization that intends to implement Continuous Improvement routines and culture is fundamental for the success of that claim. The way in which top management identifies and assumes the direction to follow largely dictates the degree of success or failure in the Continuous Improvement movement.

The vision of the organization can be understood as what the organization wants to be in an ideal future. Besides being its aspiration, the vision of an organization is also its inspiration. The vision does not have to be achievable, but more importantly it should show the way forward. It is often a goal that in practice will never be achieved, such as "zero defects" or "zero breakdowns". Organizations will never achieve for example "zero defects" or "zero breakdowns", but the fact that this theoretical ideal is assumed as direction, is extremely important to indicate the way forward for all organization staff. The advantage of the vision being unattainable is that the reference is never lost. No matter how much you improve, the direction is still there, as can be seen in the representation in Figure 3.5. The vision can also be called "the true north" in analogy of the geographic north. True north refers to the direction along the surface of the planet earth towards the North Pole. In reality, the geographic north differs from the magnetic north.

An example of a view that illustrates the impossibility of full realization is the following:

> *"Toyota will lead the way to the future of mobility, enriching lives around the world with the safest and most responsible ways of moving people. Through our commitment to quality, constant innovation and respect for the planet, we aim to exceed expectations and be rewarded with a smile"* (Toyota_Europe, 2020)

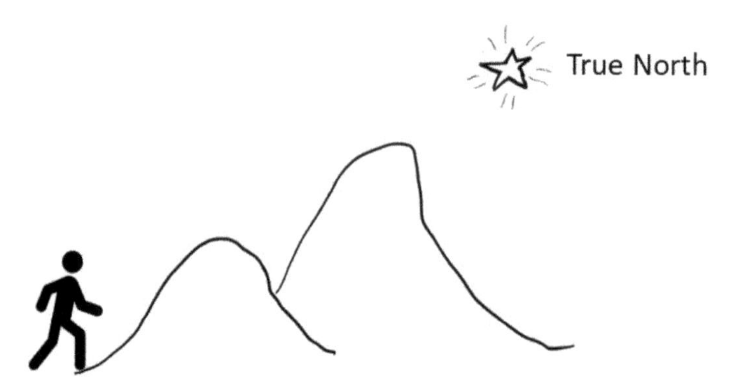

Figure 3.5 The True North.

The excerpt "*...we aim to exceed expectations and be rewarded with a smile...*" mentions something highly unlikely to be fully achieved, not least because of the subjective nature implied. For a organization to get close to such a goal, it will have to be constantly improving, and yet it will never be guaranteed to get there. That is the nature of a good vision. On the other hand, when the vision states an achievable goal, such as "to be the world's largest organization" in a particular industry, a problem arises: losing the purpose if that goal is achieved.

Other examples of vision statements that are interesting because they are unobtainable in an objective way are the following (Wright, 2018):

1. "*To fulfil dreams through the experiences of motorcycling*" (Harley-Davidson);
2. "*To delight our customers worldwide, all of the time*" (Audi);
3. "*To develop leaders who will one day make a global difference*" (Harvard University);
4. "*To make people happy*" (Disney);
5. "*Helping content creators around the world to find a global audience*" (Netflix);
6. "*The web's most convenient, secure and cost effective payments solution*" (PayPal);
7. "*To bring consumers safe, high quality foods that provide optimal nutrition*" (Nestlé);
8. "*Improving people's lives through meaningful innovation*" (Philips);
9. "*Create better everyday lives for as many people as possible*" (Ikea);

Vision statements are extremely powerful because they help the whole organization in decision making when solving problems or designing improvement actions. The solutions to be designed for improvement or to solve problems should always be a progress towards the vision. This fact facilitates decision making enormously.

3.8 The Values of an Organization

Another very relevant aspect of organizations is their values. Values are the principles that govern and condition the day-to-day behavior of the people in the organization and, desirably, the culture. These values, aligned with the vision, are also fundamental in guiding the day-to-day decision-making processes at all levels of the organization.

Examples of the values from Ikea (IKEA, 2020):

1. *"Togetherness and enthusiasm";*
2. *"Caring for people and planet";*
3. *"Cost-consciousness";*
4. *"Simplicity";*
5. *"Renew and Improve";*
6. *"Give and take responsibility".*

Examples of the values from (Business Strategy Hub, 2019)

1. *"Professional Excellence";*
2. *"Customer Oriented";*
3. *"Teamwork";*
4. *"Welcoming new challenges";*
5. *"Global Perspective".*

The vision and values defined by an organization serve to help all employees identify with the organization and contribute with their ideas and decisions so that the organization moves in the direction it has chosen.

The vision for Continuous Improvement of an organization is in turn a statement of the aspiration of Continuous Improvement itself in its contribution to the vision of the organization. An interesting vision statement for Continuous Improvement could be "improvement applied by everyone, every day".

Since we are talking about Continuous Improvement models and practices, we can say that it can and should be an extremely relevant vector in the strategy of organizations. It makes perfect sense to assume that Continuous Improvement is a strategy to improve business, increase profit, reduce costs, increase the attractiveness of products and services, improve motivation and job satisfaction of employees, and much more. On the other hand, we can also talk about the strategy for Continuous Improvement, which is to define the way in which it is put into practice. Not all organizations implement Continuous Improvement in the same way, adopting different models or different strategies.

3.9 The Culture

Culture can be defined as the kind of basic assumptions that a given group has invented, discovered or developed in its learning (both to deal with its problems of adaptation to the group's environment and internal integration of group members) and that have worked well enough to be considered valid and, therefore, to be taught to new members as the right way to perceive, think and feel in relation to these problems (Schein, 2017). The culture of a group of people, or of a society, at a particular period can be broadly defined as being its customs, its laws, its morals and its beliefs. It can also be said that culture is embodied in the way people respond to situations and how, for example, they are expected to behave in certain contexts. Culture can also be seen as the group's set of implicit informal norms, routines and decision-making patterns, habits. Peter Drucker, one of the most important modern management thinkers, left a sentence that has become very popular and depicts the power of culture, "Culture eats strategy for breakfast" (Engel, 2018). This quote makes clear the importance that an organization's culture has on its results. The culture of an organization is extremely important because, in fact, it dictates the day-to-day life of the organization. No matter how good the designed strategy is, and even if it is properly deployed by the various departments and sections, it is in fact the culture that will shape the behavior, decisions and results on the ground.

Any Continuous Improvement system must be based on a consistent culture that includes intrinsically or extrinsically the promotion of Continuous Improvement throughout the organization. The promotion of Continuous Improvement will be intrinsic when it happens naturally and will be extrinsic when it is forced by management or achieved through trade-offs. If Continuous Improvement is not closely linked to an organization's culture and strategy, then it will not be easy to establish it consistently.

As an example, Bosch's culture is based on a set of elements and principles that ensure the organization's main goal: Customer Satisfaction. All efforts are aligned with the aforementioned "True North", i.e. the point of reference, which, in this case is characterized by (Xiaoxia, 2018):

1. 100% of value added;
2. 100% of quality deliveries made;
3. Zero faults;
4. *One piece flow*.

As mentioned earlier, the state characterized by "*true north*" does not have to be attainable! It is there to set the direction, showing the direction that efforts should take. In this way, Bosch can ensure that the work contributes to bringing the organization ever closer to the desired state. Once this direction is defined, it is easier to identify the improvement actions (among several possible ones) that bring the organization closer to its "true north" and that, therefore, should be prioritized in terms of implementation.

3.10 A Higher Purpose

The identification and acceptance of a greater purpose by an entire organization is one of the strongest ingredients for that organization to achieve enormous success. It is obvious that it is not enough, but if we add the implementation of a model of excellence in organizations and the engagement of all people in the organization, then success is guaranteed. To better understand what is meant by 'a higher purpose' the classic story of the three masons is presented.

> One day, a traveler, walking on a road, came across to 3 masons working in a quarry. Each of the masons was busy cutting his own block of stone. Interested to find out what the purpose of these men's work was the traveler asked the first mason what he was doing. "As you see I am cutting this stone". Still not understanding what the masons were doing, the traveler turned to the second mason and asked the same question. The second mason replied, "I am cutting this block of stone to make sure it is perfectly square and of uniform dimensions, so that it fits exactly in its place in the wall." A little closer to figuring out what the masons were doing, but still not completely clear, he turned to the third mason and repeated the question. He seemed the happiest of the three and, when asked what he was doing, replied, "I'm building a cathedral."

This story illustrates one of the principles of the Shingo Model "Think Systemically" - seeing the whole, but also emotional alignment with the vision and purpose of the organization. The three masons were doing the same task, but each gave a different answer. Each knows how they should do their job, but what stood out about the third mason? Possibly:

- Know not only how and what to do, but also know why;
- Visualize the whole and not just its parts;
- To have a vision, a sense of being part of a coherent and grand whole;

- To have the ability to see significance in the work, beyond the obvious;
- Understand that a legacy will live on, whether in the stone of a cathedral or in the impact it may have on others.

Creating the conditions for all employees and managers to have the same way of seeing their work as the third mason should be a mission of the leadership. Although it does not directly transpire from the principles listed above associated with models that aim to achieve Excellence in organizations, it is important that there is a recognized 'higher purpose' throughout the organization. Unlike what has often been identified as a organization's purpose - profit (an example is clearly defined in the famous Goldratt's book "*The Goal*" (Goldratt & Cox, 1984) - organizations must have a 'higher purpose'. Edward Freedman, a professor at the University of Virginia Darden School of Business refers to this higher purpose using the following metaphor (Conscious Capitalism, 2021): We need to produce red blood cells to live (just as a business needs profits to live), but the purpose of life is more than producing red blood cells (just as the purpose of business is more than simply generating profit). This does not mean that generating profit can be downplayed; it is obviously vital because without profit it is not possible to pursue the higher purpose. Improving people's quality of life and health, preserving the planet, protecting the underprivileged and getting education to everyone are just possible examples of the so-called 'higher purpose'. Of course, many organizations will have some difficulty identifying a higher purpose, but it is an exercise worth doing and probably all organizations will eventually be able to identify one.

Business owners are not always able to look at their businesses in any way other than making a profit to live better lives and provide a better life for their families. There is nothing wrong with that. There are some people who have the ability and energy to be entrepreneurs and start businesses, and for that alone they should be valued. They are admirable people who should be recognized for their important role in society. They are people who create jobs and develop the economy, and even if they do not have a long-term vision or altruistic purposes, they deserve our admiration because we can guess that they have usually done hard work. The only issue is that if they do not have a higher purpose for their organizations, they will not be in a position to turn their business into something sustainable in the long term and with the potential to make a difference in the world. Many will never have any interest in this, but those who are in a stable and controlled business situation, and with some mental slack to be able to rethink everything, can try to give their business a higher purpose. In addition to achieving personal success and security for their families, is not there room to look a little further? Will they not be able to make their mark on the world? Even if they only make a difference in their region or their community? An important aspect that cannot in any way be ignored is that, in fact, when there is a clear higher purpose, it is easier for employees, managers, customers, suppliers and the community in general to align themselves to contribute their energy to that same higher purpose.

References

Bastos, A., & Sharman, C. (2018). *Strat to Action - O Método KAIZEN™ de levar a Estratégia à Prática.* Kaizen Institute.

Batt, R. (2004). Who benefits from teams? Comparing workers, supervisors, and managers. *Industrial Relations.* https://doi.org/10.1111/j.0019-8676.2004.00323.x

Benders, J., Huijgen, F., & Pekruhl, U. (2001a). Measuring group work; Findings and lessons from a European survey. *New Technology, Work and Employment.* https://doi.org/10.1111/1468-005X.00089

Benders, J., Huijgen, F., & Pekruhl, U. (2001b). Measuring group work; Findings and lessons from a European survey. *New Technology, Work and Employment.* https://doi.org/10.1111/1468-005X.00089

Business Strategy Hub. (2019). *Toyota: Vision; Mission; Values; Philosophy.* https://bstrategyhub.com/toyota-vision-mission-values-philosophy/

Cohen, S. G., & Ledford, G. E. (1994). The Effectiveness of Self-Managing Teams: A Quasi-Experiment. *Human Relations.* https://doi.org/10.1177/001872679404700102

Conscious Capitalism. (2021). *Higher Purpose.* Consciouscapitalism.Org/. https://www.consciouscapitalism.org/higher-purpose

Crabtree, S. (2013). *Worldwide, 13% of Employees Are Engaged at Work.*

Csikszentmihalyi, M. (1990). *Flow: the psychology of optimal experience.* Harper & Row.

Delarue, A., Van Hootegem, G., Procter, S., & Burridge, M. (2008). Teamworking and organizational performance: A review of survey-based research. *International Journal of Management Reviews.* https://doi.org/10.1111/j.1468-2370.2007.00227.x

Devine, F., & Bicheno, J. (2020). Creating Employee 'Pull' for Improvement: Rapid, Mass Engagement for Sustained Lean. *Lecture Notes in Networks and Systems, 122.* https://doi.org/10.1007/978-3-030-41429-0_7

Engel, J. M. (2018). Why Does Culture "Eat Strategy For Breakfast"? *Forbes.*

Fleming, D. (2018). *Why Shingo? An Overview - YouTube.* https://www.youtube.com/watch?v=j8F-4b7LVVE

Goldratt, E., & Cox, J. (1984). *The Goal: A Process of Ongoing Improvement.* North River Press.

Graban, M. (2012). *Notes from Meeting Masaaki Imai.* LeanBlog. https://www.leanblog.org/2012/06/meeting-mr-imai/

Heath, C., & Heath, D. (2010). *Switch: How to Change Things When Change Is Hard.* Broadway Books.

Hines, P., Found, P., Griffiths, G., & Harrison, R. (2011). *Staying Lean: Thriving, Not Just Surviving.* Productivity Press.

IKEA. (2020). *IKEA culture and values.* About Ikea. https://about.ikea.com/en/who-we-are/our-roots/ikea-culture-and-values

Kulkarni, S., Welekar, S., & Kedar, A. (2018). Productivity Improvement through Quality Circle: A Case Study at Calderys Nagpur. *International Journal of Mechanical and Production Engineering Research and Development (IJMPERD).* https://doi.org/10.24247/ijmperdfeb2018116

Lawler_III, E. E., & Mohrman, S. A. (1985). Quality circles after the fad. *Harvard Business Review.* https://doi.org/10.1002/cyto.990110609

Liker, J. (2004). *Toyota Way: 14 Management Principles from the World's Greatest Manufacturer.* McGraw-Hill Education.

Maslow, A. H. (1943). A theory of human motivation. *Psychological Review, 50*(4), 370–396. https://doi.org/10.1037/h0054346

McLeod, L. E. (2013). *Selling with Noble Purpose: How to Drive Revenue and Do Work That Makes You Proud.* John Wiley & Sons, Inc.

Nonaka, I. (1993). History of the quality circle. *Quality Progress, 26*(9), 81–83.

Powell, D., & Reke, E. (2019). No Lean Without Learning: Rethinking Lean Production as a Learning System. *Ameri F., Stecke K., von Cieminski G., Kiritsis D. (Eds) Advances in Production Management Systems. Production Management for the Factory of the Future. APMS 2019. IFIP Advances in Information and Communication Technology, 566.* https://doi.org/10.1007/978-3-030-30000-5_8

Revans, R. W. (1980). *Action learning: new techniques for management.* Blond & Briggs.

Rother, M. (2010). *Toyota Kata: Managing People for Improvement, Adaptiveness and Superior Results.* McGraw-Hill Education.

Sagi, S. (2015). "Ringi System" The Decision Making Process in Japanese Management Systems: An Overview. *International Journal of Management and Humanities.*

Schein, E. (2017). Organizational Culture and Leadership Organizational Culture and Leadership. In *Wiley & Sons, Inc.*

Senge, P. M. (2006). *The fifth discipline: the art and practice of the learning organization.* Doubleday.

Stewart, J. (2011). *The Toyota Kaizen Continuum: A Practical Guide to Implementing Lean*. Productivity Press.

Sugimori, Y., Kusunoki, K., Cho, F., & Uchikawa, S. (1977). Toyota APAGAR. *International Journal of Production Research*, *15*(6), 553–564. https://doi.org/10.1080/00207547708943149

Toyota_Europe. (2020). *Toyota Global Vision and guiding principles*. Toyota_Europe Web Page. https://www.toyota-europe.com/world-of-toyota/this-is-toyota/toyota-global-vision

Womack, J., Jones, D., & Roos, D. (1990). *The machine that changed the world*. Free Press.

Womack, J. P., & Jones, D. T. (1996). Lean Thinking by Womack and Jones. *Review Literature And Arts Of The Americas*, *November*, 5.

Wright, T. (2018). *Vision Statement Examples*. https://www.cascade.app/blog/examples-good-vision-statements

Xiaoxia, R. (2018). *Bosch Connected Industry*. Leanchine.Net.Cn. http://www.leanchina.net.cn/wp-content/uploads/2018/07/Bosch-Connected-Industry-Amanda-Ren.pdf

4

The Visible Side

This chapter is focused on some of the main visible aspect of Toyota Excellence and what became in fact the main target of interest for the scientific and business community around the world when TPS started to raise global curiosity. We will therefore put special emphasis on the technical side associated with some of the principles of the 3 reference models we have covered in this book (Lean Philosophy, Shingo Model and Toyota Way). The principles of continuous flow production and demand-driven flow will have particular relevance in this chapter, not only because they represent a disruptive paradigm shift, but also because they attract curiosity and also because they are often recognized as counterintuitive.

There are several principles, present in excellence models, which clearly focus on the more technical/visible side of organization and operations management. This chapter will present and discuss the principles shown in Table 4.1, which are essentially shared by the three chosen reference models that seek Excellence in organizations.

One of the greatest contributions to improving the competitiveness of organizations, originating with Henry Ford's assembly line and later developed/explored more fully in the *Toyota Production System* (TPS), was the concept of making production flow. The idea that materials should move from one process to another as quickly as possible, although at first sight it may seem a simple and even intuitive idea, is actually not quite so. Before developing this subject, it is important to clarify an important aspect; a traditional assembly line undoubtedly guarantees fluidity, but it only achieves this because it only produces a single type of product (mass production). The great paradigm shift proposed by TPS, which is one of its greatest contributions/

Table 4.1 Principles on the more technical/visible side of excellence in organizations.

Shingo Model	Toyota Way	Lean
Improve Flow & Pull	**Principle 2**. Create a continuous process flow to bring problems to the surface. **Principle 3**. Use "pull" systems to avoid overproduction. **Principle 4**. Level out the workload (*Heijunka*). (Work like the tortoise, not the hare).	**Principle 3**. Flow **Principle 4**. Pull
Focus on Process	**Principle 6**. Standardized tasks and processes are the foundation for continuous improvement and employee empowerment. **Principle 7**. Use visual control so no problems are hidden. **Principle 8**. Use only reliable, thoroughly tested technology that serves your people and processes.	**Principle 2**. Identify the Value Stream

innovations, is to achieve fluidity even when producing different types of products (diversity).

This concept of making things flow is extremely powerful and not only applies to physical materials and products. It also applies to information processing or services in general, and can even be used in our personal life (e.g. how we let a lot of unfinished business pile up or not). By itself, this subject of personal management would justify a book - in fact, there are already several books published on this subject, one of them being the very interesting "*Factory of One*" (Markovitz, 2011). Overall, keeping materials, services and issues waiting for processing, i.e. stagnant, results in more work, stress and anxiety.

4.1 Some Key Performance Indicators in Production

The Key Performance Indicators (KPIs) adopted and monitored by any given production unit mirror the contribution of that production unit to the organization's vision and strategy. The KPIs chosen are the materialization of the organization's strategy and determine/condition the decisions to be made on a daily basis. In terms of operations, each of the operational teams is conditioned to pursue objectives (values) for their own KPIs but these must be aligned with the objectives defined for the organization's strategy. As an example, to clarify the alignment or non-alignment of indicators, let us

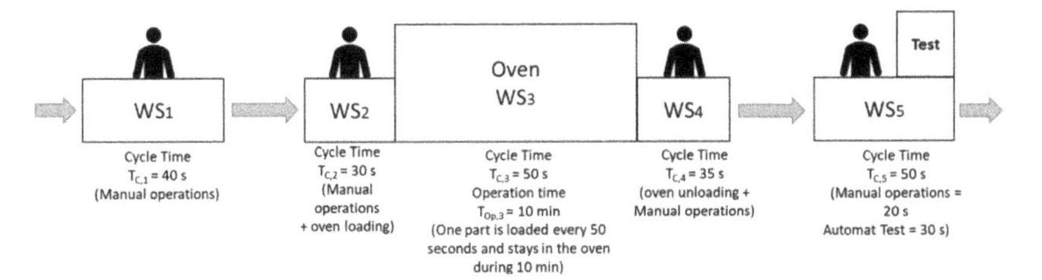

Figure 4.1 A small sequential production unit.

imagine that one of a organization's KPIs is the defect rate (with the goal of reducing it to zero) but each operational team only pursues productivity. In this case there is no alignment of the organization's indicators with the ones pursued by the operational teams. This type of misalignment is not so rare, nor is the dispersion of objectives, as well as the definition of contradictory objectives within an organization.

Although not completely associated with production fluidity, it is important to present some relevant production indicators since it will be necessary to refer to them throughout this chapter. To facilitate the description of each indicator, we decided to use as an example a small production line composed by 5 workstations (Figure 4.1).

First let's describe the concept of **Cycle Time** (T_C). The cycle time of a workstation is the period of time that elapses between two consecutive units produced by that workstation. In other words, T_C indicates how often a part leaves the workstation. For example, the workstation WS_1 delivers a part every 40 seconds ($T_{C,1} = 40s$). The next concept is the **Time for the operations** (T_{Op}), or processing time, which is the time during which a part unit is being processed. The cycle time ($T_{C,1}$) of workstation WS_1 is equal to the **Operation time** ($T_{Op,1}$) because it is a manual workstation, however, this is not always the case. If we look at workstation WS_3 (oven), although it can finish a part every 50 seconds, each part is inside the oven being processed during 10 minutes. Thus, in this case, the cycle time is 50 seconds $T_{C,3} = 40s$ while the processing time is 10 minutes ($T_{P,3} = 10\ min$).

As mentioned, in the first workstation only manual operations are performed. The same occurs at workstation WS_2 where the operator starts by applying glue to the part and then places it in the oven (workstation WS_3). In the oven there is a conveyor belt that moves the parts from the entrance until the other end of the oven. The operator of the workstation WS_2 finishes a part every 30 seconds but can only load one part into the oven every 50 seconds, which means that every 50 second cycle has 20 seconds of inactivity. Although one part is loaded every 50 seconds, each one of them spends 10 minutes in the oven, in a glue drying process ($T_{Op,3}$ = 10 *min*) and leaves the oven, obviously, one part every 50 seconds ($T_{C,3}$ = 50*s*). At workstation WS_4 the operator unloads the parts from the oven and removes excess glue from them, spending 35 seconds in these operations ($T_{C,4} = T_{Op,4}$ = 35*s*). At workstation WS_5 the operator places the part in an automatic test equipment, requiring 20 seconds, and then waits for the automatic cycle of 30 seconds to take place.

Although each workstation (WS_i) has its cycle time ($T_{C,i}$), the line as a whole has its cycle time ($T_{C,Line}$). The cycle time of a line is dictated by the workstation with the longest cycle time (bottleneck), i.e.:

$$T_{C,Line} = \max (T_{C,i})$$

In this case, we then have $T_{C,Line}$ = max (40,30,50,35,50) = 50*s*, which is the same as saying that every 50 seconds, if no unforeseen events occur, the line delivers one part. This is a very important indicator of the expected capacity of the line.

Another indicator that derives from the cycle time (T_C) is the **Production Rate** (R_P), also known as production capacity, output or throughput[1], which indicates the quantity of parts/products/articles/units that a processor, or set of processors, can produce per unit of time, being given by[2]:

$$R_P = \frac{1}{T_C}$$

In the case of the line mentioned above, we have $R_{P,Line}$ = 1/50 = 0.02 *parts/s* = 72 *parts/h*.

This is the theoretical production rate because in fact, in the general case, a line never produces exactly at that rate due to several reasons such as breakdowns, small breaks, material jams, etc. In general terms, **Productivity** (*Pr*) is a performance indicator that indicates how efficiently resources are used to create value. We can talk about the productivity of a country or region, which is obtained by dividing the Gross Domestic Product of that country or region by the number of employees or by the number of hours worked in the same country or region. In the present context, when we talk

---

[1] "Throughput" is the term adopted by the "father" of *Theory Of Constraints*, Eliyahu Goldratt.

[2] In dimensional terms this equation would only be represented in parts/second if the cycle time were in seconds/part. However, when being in seconds/part it could not be designated as cycle time because it is not a time, but a period (inverse of a frequency). In other words, the assumed incorrectness is to use the cycle time in this production rate formula (strictly it is not the cycle time in seconds, but the period in seconds/part).

about productivity we are referring to the quotient between the production rate (R_p), ie the number of parts produced in a given period of time, and the amount of labor (n) needed to produce these parts (e.g. number of workers). Thus, in a given period, labor productivity can be described as the average contribution of each employee in the production of parts or products in that period. Typically, this type of productivity is presented in terms of units produced per hour and person, given by:

$$Pr = \frac{R_P}{n}$$

For the case of the production unit that we have been analyzing, the expected productivity would then be Pr_{line} = 72/4 = 18 parts/man.h, i.e., it would be expected that, every hour in average, each worker would produce 18 parts. Although this is the theoretical productivity expected for the line in question, at the end of each day, or shift, the actual productivity can be evaluated. There are many reasons why the quantity of parts produced is not the same as expected, namely: machine breakdowns, variability in processing times, production of defects, jamming of components, material shortages, human errors, incidents and accidents, wear of equipment and tools, etc.

When, for a given period of time (T), it is intended to evaluate the productivity (Pr) of a production unit with n workers, based on the quantity of products/parts that this unit has produced (Q) in that period, and also considering a certain quantity of products/parts that were produced with defects (Q_d), the following equation must be used:

$$Pr = \frac{Q - Q_D}{n \times T}$$

As an example, assuming that the production line under study produced only 460 parts in a shift, of which 10 did not pass quality control, and even that the 8-hour shift had two 10-minute breaks. Thus, the productivity of this shift would be:

$$Pr = \frac{460 - 10}{4 \times \left(\frac{480 - 2 \times 10}{60} \right)} = 14.7 \, parts \, / \, man.h$$

Four people working for 460 minutes represent 30.7 man.hour in terms of manpower. Each person contributed, on average, with the production of 14.67 parts for each hour of work. This value is lower than the expected value for the line which was 18 parts per person in one hour.

Line Efficiency (E) is one performance indicator that refers to the intrinsic characteristics (how it was designed) of a production or assembly line with several workstation and with its line cycle time $T_{C,Line}$. This performance indicator evaluates in a certain way how similar are the cycle times of all its all workstations ($T_{C,i}$). Line efficiency is given by the following equation:

$$E = \frac{\sum_{i=1}^{n_{WS}} T_{C,i}}{n_{WS} \times T_{C,Line}}$$

(*Being* n_{WS} *the number of workstations*)

A line will only be 100% efficient if all workstations that comprise it have the same cycle time. Thus, for the case represented in Figure 4.1, the line efficiency is:

$$E = \frac{\sum_{i=1}^{5} T_{C,i}}{5 \times 50} = \frac{40+30+50+35+50}{250} = 82\%$$

Given that workstations can have very expensive resources, it is easy to understand the importance of using these resources well, and, in this sense, trying to ensure that the line efficiency is as high as possible.

From the point of view of the human resources needed on the line, since these are valuable and costly resources, it naturally makes sense to consider an indicator related to the use of labor. This aspect is particularly relevant in regions with high salaries. In fact, when designing a line it is important to predict the extent to which labor is being used in value adding operations (transformation on the product/part). This indicator is the **Labor Utilization Rate** (R_{LU}) and, for a line with cycle time $T_{C,Line}$ and with n workers, each worker with its operating time , is given by the following equation:

$$R_{LU} = \frac{\sum_{i=1}^{n} T_{Op,i}}{n \times T_{C,Line}}$$

Using this equation regarding the production line represented in Figure 4.1, the result is:

$$R_{LU} = \frac{\sum_{i=1}^{4} T_{Op,i}}{4 \times 50} = \frac{40+30+35+20}{4 \times 50} = 62.5\%$$

Note that although the cycle time of workstation is 50 seconds ($T_{C,5} = 50s$), the operator only carries out operations for 20 seconds ($T_{Op,5} = 20s$).

In practical terms, operators in this production line use, in average, only 62.5% of their time in value adding operations on the product. This is actually the average expected value for the line as it is designed, certainly varying from day to day.

An alternative way to calculate the rate of use of labor in a given period of time (T), although derived from the one presented above, considers the total quantity of products/parts and the quantity of defective products/parts produced in that same period of time (Qe Q_D, respectively) and is given by:

$$R_{LU} = \frac{\left(Q - Q_D\right) \times \sum_{i=1}^{n} T_{Op,i}}{n \times T}$$

If we again assume that, in one shift, the production line depicted in Figure 4.1 produced only 460 parts, 10 of which failed quality control, and even though the 8-hour shift had two 10-minute breaks, then the utilization rate of the workforce for this shift will be:

$$R_{LU} = \frac{\left(460 - 10\right) \times \left(40 + 30 + 35 + 20\right)}{4 \times \left(8 \times 60 - 2 \times 10\right) \times 60} = 50.9\%$$

This performance indicator measurement is widely used in organizations where the cost of labor plays a major role in the product price ("labor-intensive" organizations), namely organizations in the footwear production, clothing industry and major assembly sectors. In a large part of these types of factories, this indicator is sometimes referred to as "productivity" or "efficiency".

Although not widely used in many small and medium organizations, a key indicator of fluidity issues in production is the amount of materials in the process of being manufactured. Its most common designation is WIP (Work-In-Process). The WIP associated with a production process is the quantity of products that are being processed and/or waiting to be processed in that same process. One of the problems with this indicator is that, in general, its value varies depending on the day and also throughout the day, and it is therefore necessary to define and understand its meaning well. When this performance indicator is used in the calculation of other indicators, it is advisable to be careful with the assumptions, as will be clarified below.

Observing the production line represented in Figure 4.2, we verify that it is practically identical to the one illustrated in Figure 4.1. The difference lies in the inclusion to each process/workstation, of a triangular symbol with

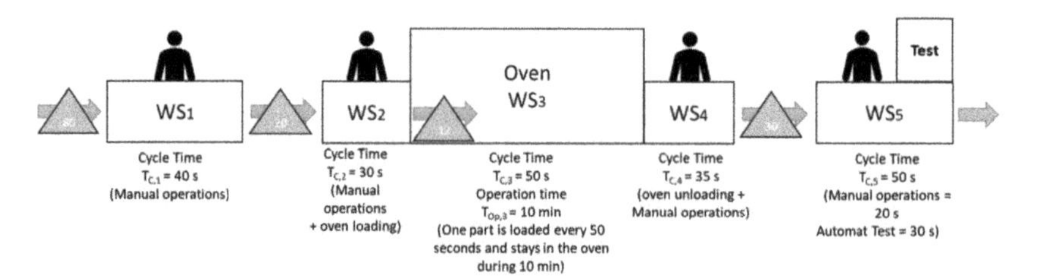

Figure 4.2 The production unit with WIP.

a numerical value. This triangle represents stock (in this case *WIP*) and is adopted in this book as it is a symbol already used with some popularity, especially in VSM (*Value Stream Mapping*)[3] tool. The value in each triangle represents the quantity of products/parts/items that are waiting to be processed and/or to be processed at the associated workstation. In the situation illustrated in Figure 4.2, workstation WS_1 has a WIP of 80 parts (WIP_1 = 80 *parts*), one of which is probably currently being processed. As can be seen, all *WIP* associated with the oven (WIP_3 = 12 parts) is being processed (i.e. it is inside the oven) because the operator of the workstation WS_2 loads the oven with one part each 50 seconds (i.e. there is no inventory between its workstation and the oven). The oven *WIP* is 12 parts because that is exactly the maximum capacity of that equipment. As you would expect, the total *WIP* of the line (WIP_{Line}) is the sum of the WIP of each of the workstations that compose it (WIP_i), that is:

$$WIP_{Linha} = \sum_{i=1}^{n_{WS}} WIP_i$$

Thus, in this case (Figure 4.2), we have WIP_{Line} = 80 + 20 + 12 + 30 = 142 *parts*.

4.2 The Concept of Flow

The amount of *WIP* that exists in a production unit at a given instant is an indicator of the fluidity of that same production unit at that instant. The higher the *WIP* value, the lower the fluidity and vice versa. So, what is the fluidity of the production unit that we have been analyzing (Figure 4.2)? One way to

[3] VSM (Value Stream Mapping) was presented in two famous books published at the turn of the millennium: the book "Learning to See" by Rother & Shook (1999) and the book "Seeing the Whole" by Jones & Womack (2002). A brief description is given in the annex.

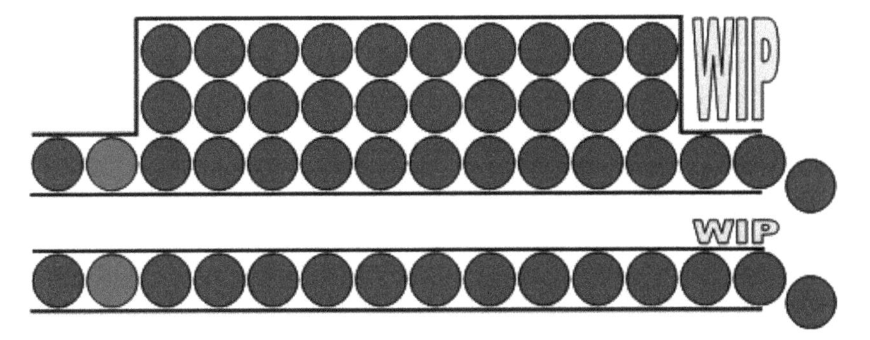

Figure 4.3 Effect of WIP in the flow speed.

evaluate it is to determine the time it takes for a part to go through the entire process, that is, the amount of time that elapses between the instant when that part arrives at queue or buffer upstream the workstation , joining the 80 parts already there, and the instant when it is ready at the last workstation (WS_5). It is important to note that the parts will have to wait in each queue (each triangle represented in Figure 4.2), ideally according to the FIFO (*First In, First Out*) discipline. We choose to call this time **Throughput Time** (T_T) but sometimes people refer it as *lead time*. The term "throughput" is used here because it is associated with the idea of "traversing" or "going through" the entire system, whether that system is an entire plant or just a part of it.

The flow of materials in production can be compared to the flow of a liquid in a pipeline and this analogy can be very effective in understanding the concept of production flow. If we decrease the cross section of a pipeline, in order for the flow to be maintained, the fluid velocity has to increase. Of course, as you increase the velocity, each particle of fluid will travel faster through the pipeline. A very similar behavior also exists in production systems. In this analogy, the quantity of "balls" in the pipeline represents, in some way, the quantity of products/parts/material in progress, that is, the WIP (Figure 4.3). Therefore, for the same production rate, the higher the *WIP*, the longer the time each product takes in the system and vice versa.

This theory was formally introduced by John Little (Little, 1961) and became known as Little's Law. According to this law, the throughput time T_T of a production system with a cycle time T_C and a quantity of products/parts in *WIP* manufacturing process is given by:

$$T_T = WIP \times T_C$$

Thus, for the production line under study (Figure 4.2), the Throughput time is:

$$T_T = 142 \times 50 = 7100\ sec = 118.3\ min$$

While cycle times are easily understood by most people, as well as operation times or processing times, the same is not true when we are talking about the concept of throughput time. Intuition (or "common sense") usually does not help most people very much in this matter. As a rule, managers and supervisors of production units are quite optimistic with regard to the value of throughput time on their production units, because they do not understand very well the real impact of inventories. Although it is not the only one, this is one of the reasons why most organizations have great difficulty in meeting the delivery deadlines agreed with their customers; and even when they do manage to deliver it is, in large part, at the expense of additional effort when the due date is approaching.

This is also why many managers use urgency level identification schemes for certain parts/batches so that they can bypass the queues (thus disrespecting FIFO discipline) and get to customers faster. They forget, however, that with this approach they are making the parts/batches that were already waiting and that are passed over, have to wait even longer. And then the question may be asked: if these parts/lots could be overrun, why were they waiting there? The evidence shows that, after all, the outdated parts/lots could have been brought later to queue. Well, this type of discussion is usually complex, with exchanges of arguments that we know very well, especially on the part of those who are not yet fully aware of the importance of flow; but the truth is that most organizations need to improve flow to become more competitive.

Regarding the understanding of production, it is important to note that our intuition and common sense can lead us to wrong paths and that, before we can effectively use that intuition and common sense, we must understand as well as possible the behavior of production flows. The principles of scientific thinking play a very important role in this context because they allow us to avoid the possible errors caused by intuition. There is nothing really special about the behavior of material flows, in fact it is quite simple, but wrong interpretations usually lead to inefficiency and loss of competitiveness. What will be explained below is nothing new, but we know that it is still not clearly understood by some professionals in this area of knowledge.

A key piece of knowledge regarding the importance of flow is that the less time products are idle waiting for some operation to happen on them, the better the overall performance of the organization. This idea is not new;

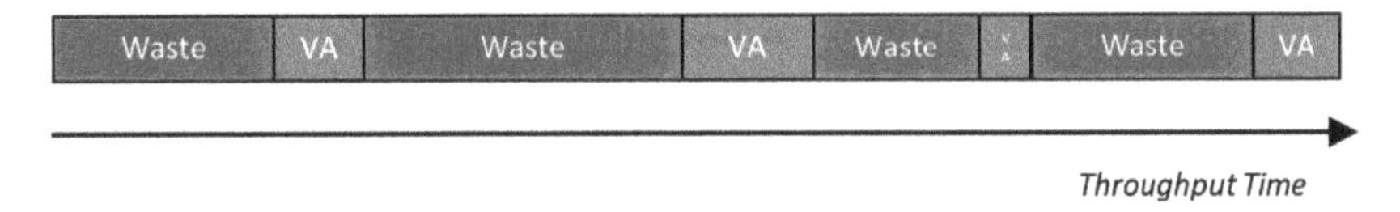

Figure 4.4 Value Adding moments through the throughput time.

in fact Henry Ford already mentioned it a long time ago, more precisely in 1926, saying the following (Alukal & Manos, 2006):

> *"One of the most noteworthy accomplishments in keeping the price of Ford products low is the gradual shortening of the production cycle. The longer an article is in the process of manufacture and the more it is moved about, the greater is its ultimate cost."*

The absolute value of what can be considered a "good" throughout time varies depending on the type of production. It can be expected that the throughout time for an assembly line for small boats will be higher than the throughout time for an assembly line for car radios, mainly because the operation times in the first case will certainly be longer. For this reason, another indicator has been created that is intended to be more universal, not dependent on operation times.

It is often indicated in several publications, such as Beecroft, Duffy, & Moran (2003) and PPDT (1998) that 95% of the throughput time is spent in activities without any value being added to the products (red areas in Figure 4.4). In other words, this means that products are standing still, with nothing happening to them, for 95% of the time they spend in the production system. Moreover, this figure is given as the benchmark for world-class organizations, meaning that for most organizations this percentage will be even higher.

A performance indicator used to measure this relationship between the time spent on value added activities (i.e. the sum of the operation times ($T_{Op,i}$), for m operations) and the throughput time T_T, is called the **Value Added Ratio** (R_{VA}) and is calculated as follows:

$$R_{VA} = \frac{\sum_{i=1}^{m} T_{Op,i}}{T_T}$$

Thus, for the production line under study (Figure 4.2), the value added ratio would be:

$$R_{VA} = \frac{\sum_{i=1}^{5} T_{Op,i}}{118.3 \times 60} = \frac{40 + 30 + 10 \times 60 + 35 + 50}{118.3 \times 60} = 10.6\%$$

The higher the R_{VA}, the greater the fluidity and the more "*Lean*" the production is, i.e. the products wait less time to go through the various production steps. Values for R_{VA} above 5% are considered good, according to what was previously mentioned. The objective should always be to increase the R_{VA}, reaching the maximum value when the ideal fluidity is achieved, that is, the so-called *One-Piece-Flow* production. There is, however, an important detail regarding the measurement of this indicator that is important to mention: the selection of the beginning and the end or the production unit to analyze in measuring the throughput time. Imagine, for example, the following two cases: (i) the throughput time starts to be measured when the materials are delivered by the supplier to the raw materials warehouse and ends when the products are shipped to the customers, and (ii) the time starts to be measured when the materials are picked up from the raw materials warehouse and ends when the products are shipped to the finished goods warehouse. Evidently, measurement according to option (i) leads to higher throughput times.

The main message that this indicator intends to convey is that the overwhelming majority of the time that products spend in production units is spent waiting (as inventory) for the next production process. In the several measurements that we made in dozens of organizations, the value of the measured Value Added Ratio (R_{VA}) was always lower than the 5% used as a reference for world-class organizations. In a large number of them the value for R_{VA} was actually below 1%. In services, the figure is even worse. Even in the cases in which the measurement of the throughput time was carried out considering option (ii) presented in the previous paragraph, the value for R_{VA} obtained was always well below 5%.

Note that the value of R_{VA} we obtained for the production unit under study (Figure 4.2) is much higher than the reference value of 5% but that happens due to the existence of an oven with a processing time of 10 minutes. Moreover, we are only counting the throughput time for the line, i.e., ignoring the time that the parts spent as inventory before arriving to the first workstation as well as the time that the parts will take until they are dispatched.

4.3 Allocate Resources where the Effect on Fluidity is Greatest

Making things flow is a very important concept and principle that we come across every day in industrial organizations in general, but also in other contexts. The context we want to give you here as an example is the context of a bar or cafeteria (Figure 4.5) which can be found in schools and universities campus but also in other similar contexts.

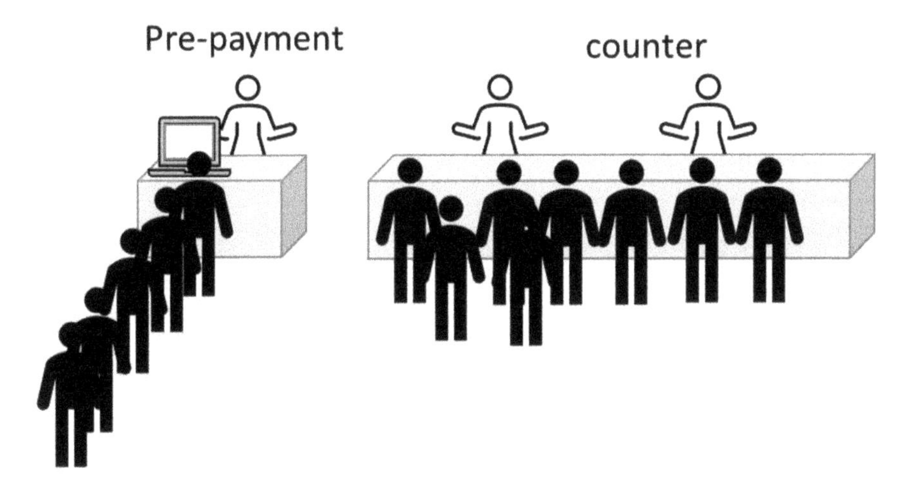

Figure 4.5 Example of a cafeteria in the university campus.

In this type of cafeteria customers will first have to go to a queue to pay for what they want (pre-payment) and then move to a counter where they wait to be served. It is not uncommon to find a scenario similar to the one depicted in Figure 4.5 and the reader might even think that this is acceptable. Now let us look at this scenario with the concept of flow in mind. The pre-payment clerk is not contributing to the flow; he is just moving the customers from the pre-payment queue to the waiting queue in the counter, i.e. he or she is just moving the customers from one queue to another. If that employee were to fill the orders for the counter, then he would actually contribute to more customers being served, thus reducing the amount of customers waiting. Only when the number of customers waiting at the counter is very low (e.g., one for each employee, possibly plus one or two more as a safety margin), would an employee advance more customers from the pre-payment to the bar, but always in small numbers to maintain the flow. Maybe the reader already noticed that in the most popular hamburger restaurants, like McDonalds and Burger King, they have a different approach. In these type of fast food restaurants, the principles of pull production are applied with great rigor and effectiveness. It is usually verified that the pre-payment is performed by the same employee that processes the order, and that the number of customers waiting to be served is always kept relatively under control.

Besides the throughput time being directly proportional to the WIP, there are other detrimental effects to the overall performance of the production unit. The higher the WIP, the more space it takes up and, obviously, the greater the need for space between equipment. More widely spaced equipment means that people and materials will have to travel greater distances. Having a lot of WIP will also make it harder to find the parts that are needed at any given time. A lot of other waste is created by having WIP and that varies from case to case.

> In a company in the Braga area (city in the north of Portugal) dedicated to the manufacture of metal parts in small batches something bizarre usually happens. Since they follow the traditional pushed production way, many production batches are naturally waiting between processes, everywhere in the shop floor. At each moment, the batch to be processed is very frequently chosen according to customer's pressure level. For this reason, some batches remained in the queues for extremely long time, because others batches are more urgent, resulting in long delivery delays and poor customer satisfaction. This is already bad enough but a serious situation sometimes occurs when customer complain about a delayed batch, long overdue, and no one is able to find it in the shop floor, no matter how much they search for it. To solve the problem a new production order is issued with "extremely high" urgency, to overcome the problem. Inevitably, someone will later find the batches that had been lost somewhere in the process, and would have to scrap it (with the inherent losses). This is an extreme example of losses due to excess WIP, but many others can be found.

Going back to the cafeteria scenario mentioned above, the mode of operation adopted in that case gives rise to waste of a different nature. Let us give some examples:

i. Both the pre-payment employee and the employee serving the order will have to know what the order is. Very often, the customer will have to explain to both the details of what he/she wants. In other words, two employees have to spend some time doing exactly the same task (understanding what the customer wants) - obviously this configures a situation of task duplication (redundancy), impacting negatively on the performance of the whole process;

ii. Sometimes, when the customer is about to be served (after having pre-paid and waited at the service counter), the product he/she wanted is already sold out (although it is possible to avoid this type of situation in this operating mode, it increases the complexity of management);

iii. With some frequency, poor communication may occur during the pre-payment process (misinterpretation of the customer's requests) and this gives rise to confusion and rework at customer service. In fact, it may happen that the counter staff prepares an order that does not correspond to what the customer wants (defect), which needs to be corrected (rework) and, in the limit, may even lead to the loss of products..

In general, the importance of the principle of focus on flow comes from the result that its application entails. The adoption of this principle conditions behavior and decision-making, and guarantees consistency in the result. If the principle of flow is always taken into account when making decisions on resource allocation (at each moment), the result will be better (as it would be in the case of the cafeteria).

4.4 Bringing Problems to the Surface

Besides the more or less obvious advantages, that one can extrapolate from the greater fluidity of production there is a very interesting one related to an aspect explicitly mentioned in the description of the *Toyota Way* principle: "Creating a continuous process flow to bring problems to the surface". We believe that many people will not understand the fullness of this message. Is bringing problems to the surface a good thing? Do we like being told there are problems to solve? Somehow trying to answer this last question, we admit that almost everyone will know a typical sentence like "*I don't want you to bring me problems; I want you to bring me solutions*". The truth is that our nature makes us not like problems to be brought to us; nobody likes another problem to solve, especially on a day that is not going well.

On the other hand, the reader may also agree that it is not good if the employees of your organization are hiding the problems "under the carpet". In this context, most of us do not want problems to be hidden from us, but in fact, we do not like them to be shown to us either (although we know they exist); so there is a contradictory situation here. Regardless of the emotional relationship we have with the explanation of the problems that we know (or not) exist; the truth is that there is no gain in hiding these problems. In fact, and in general terms, hiding and ignoring problems is not beneficial for organizations, as we are postponing their resolution. So bringing problems to the surface will clearly contribute to continuous improvement and the pursuit of excellence in organizations.

In the production physics point of view, higher fluidity means lower stock of materials (WIP) along the value chain. On the other hand, the lower

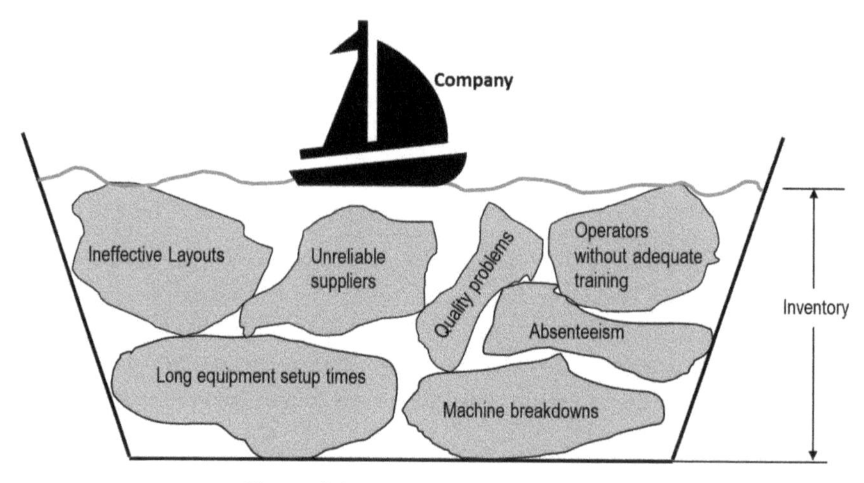

Figure 4.6 Analogy of the sailing boat.

the inventory level, the greater the risk of the organization running into problems it already has, but which are in fact hidden (see analogy in Figure 4.6). This analogy is actually a classic presented decades ago to illustrate the relationship between potential problems (represented as rocks under water) and the existing stock level (raw materials, work in progress and finished product). If the stock level is high enough, the organization will not feel the consequences of its problems. The key issue is that maintaining high stock levels involves a lot of work since human resources have to manage them, move them around, travel longer distances when going around them, etc. Moreover keeping stocks costs a lot of money (cost of production, cost of ownership, cost of capital, etc.). Finally, keeping stocks brings other kinds of inconveniences such as occupied space, risk of deterioration/obsolescence, etc., and in fact - and this is a very important element – the stock does not solve the underlying problems.

If a production unit is working in continuous flow, and let us imagine continuous flow at its maximum exponent, which is the so-called one-piece-flow production[4], the stock level will be low. In such conditions, the occurrence of any problem will immediately become visible and negatively affect the production unit (the boat will run aground on the rocks). For example, if

[4] One-piece-flow in a production unit is a type of production where only one unit of the part is allowed between each process in that production unit. Under normal operating conditions, it is not allowed to have parts waiting to be processed. When a part leaves the upstream process to the next downstream process, the downstream process is free to receive and process that part.

defects occur, having too many parts/products (overproduction, high stock levels) is comfortable, because there will always be enough to replace those parts/products that have been produced with a defect (and so the real problem - the defects – does not need to be really addressed). Another example is the case of stock kept because suppliers are unreliable (as they are late with deliveries, we tend to order too much and before it is necessary, thus bringing stock indoors because of a problem, which is not ours). For the traditional manager it is naturally more comfortable to see all the employees working, even if there is no flow (materials stored everywhere), than to see everything stopped because something went wrong. If one piece of equipment breaks down, all the others continue to work without a problem and everything seems fine since all the other pieces of equipment continue to operate because there is stock upstream of all of them. The operator that was operating that equipment even goes to another equipment and everything is fine, but, in reality, we only hide the problems. This change in the way of looking at production is a new paradigm created by Toyota.

Let us consider another case that shows the difference in paradigm of TPS in relation to the traditional approach. In traditional assembly lines, when a defective product is detected at one of its workstations, that product is removed from the line in order to be repaired later. After that the line continues to work normally. In the TPS approach, the line stops until the problem is solved and only then, the line starts working again. Any worker has the power to stop the whole line when a defect is detected. This way of managing an assembly line is counterintuitive for a large proportion of managers, as they believe that it would be better to remove the faulty product and keep the line running. The advantage of the TPS approach is that there is a huge focus on fixing the problems so that the line can start running again. In the beginning, the line will stop many times, but as the problems are solved, this will happen less and less, until a point is reached when the line stoppages become very rare.

It is important to note that you cannot increase the fluidity (reduce stocks/WIP) more than the system can support. In many cases, it is possible to increase the fluidity significantly without any problem. However, normally, in order to achieve this significant increase in fluidity without unpleasant consequences, it is necessary to find a balance between the stock level and the nature and state of the processes concerned (as shown in the analogy in Figure 4.6). It is necessary to introduce some changes in the processes, such as: reducing the setup time of some equipment, increasing the reliability of the equipment, increasing the versatility of the operators, introducing flow control mechanisms, etc., so that the fluidity can increase in a sustained manner.

The great lesson of these principles of fluidity (focus on flow) is that one should continuously seek improvement solutions that lead to greater fluidity of products throughout the various processes. All decisions, namely the purchase of new equipment, change of technology, increase in production capacity, etc., should always be conditioned by these principles of fluidity. If it is necessary to purchase a new piece of equipment to replace one that is at the end of its life (or because a new technology has emerged), then the decision among the various options should not be based on the cost of production per piece that this equipment will provide, but rather on the impact it will have on production flow.

4.5 Demand Pulled Flow

To achieve controlled flow in production it is necessary to create mechanisms that limit and control the amounts of WIP throughout the manufacturing value stream. Many professionals in this field use the term "Tense Flow" to refer precisely to this need to keep WIP controlled and at minimum values throughout the process (including shipping). This term has a very interesting physical interpretation when imagined in the production process chain: it suggests that the flow can "burst" if there is a weak point. In other words, if something goes wrong in any process, all production stops.

From the various mechanisms that can be used to control WIP, we will mention only the main ones. But before that, it is important to introduce the concept of **Takt Time** (T_T). Apparently, the term "*Takt*" was taken from the German word *Taktzeit*, which means cycle time, and was brought to Japan by German engineers during the 1930s. In any case, it is a very popular term to refer to the cadence of the market. To put it more precisely, Takt Time can be described as "the average time between two consecutive units (of the product) requested by the market considering the time available for production": So, for a production unit that has an available production time and is subject to a market demand D, the Takt Time is given by:

$$T_{Tk} = \frac{T_P}{D}$$

The idea is that once it is known the Takt Time value for the existing and accepted market, the organization can organize its production to produce

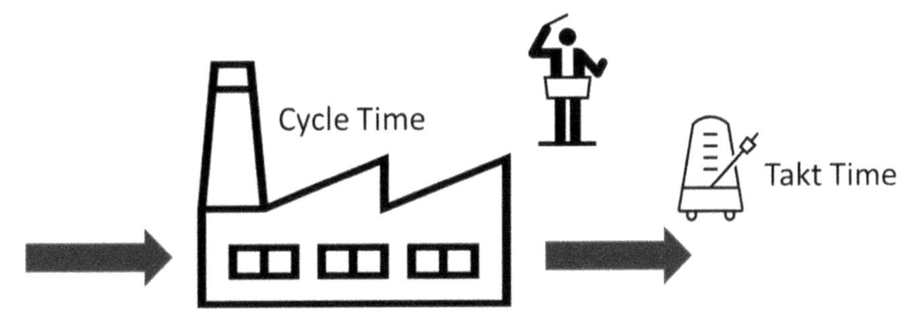

Figure 4.7 Takt time and cycle time.

a product at that same rate (imposed by the cycle time), as illustrated in Figure 4.7. This is the essence of pull flow production. Of course, this is a theoretical ideal, since in real context it is not easy to achieve. In fact, cases where the market wants one unit of the product at a time are rare. Normally, customers place orders for batches of products and that is how the transport is carried out between supplier and customer. Another illusion occurs with regard to market demand. In fact, organizations transform uncontrollable customer demand into master production plans that are suited to their capacity. If, in a given week, demand is much higher than capacity, then the organization has to find solutions such as increasing capacity by overtime or subcontracting, or negotiating with the customer delivery times for the following week or staggered deliveries so that production is as stable as possible. This means that the demand considered above in the *Takt Time* equation is not exactly the market demand, but the demand transformed by master production planning decisions.

The takt time assumed by the company affects its entire organization; since it forces it to achieve cycle times throughout all production processes in the value chain that allow it to respond effectively to that takt time. Within the organization, this same principle of cycle time adjustment to the takt time extends throughout the entire production chain of processes, i.e., between all customer processes and supplier processes.

There are many books and publications available on the internet that deal with pull production systems, namely *kanban*, *two-bin systems*, *CONWIP* (Constant Work In Process) and *POLCA* (Paired-Cell Overlapping Loops of Cards with Authorization), and for this reason we will not present them here. However, we will try to contribute with some aspects that are relevant but usually not mentioned in these publications.

Let us start with a scenario with a single supplier process and a single customer process (Figure 4.8). To define how WIP should be controlled, we

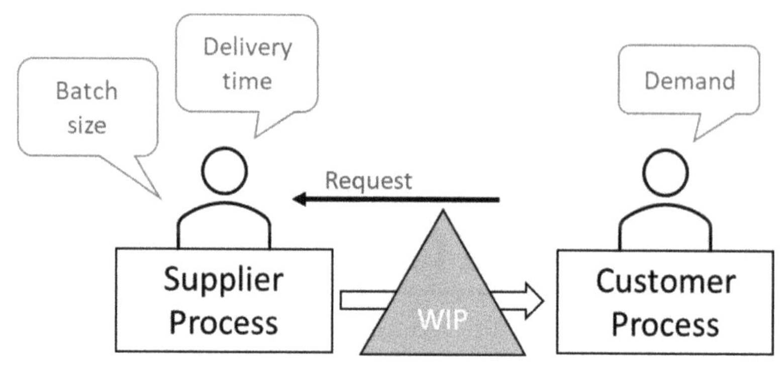

Figure 4.8 WIP sizing.

need to know the supplier delivery time (*DT*), the lot size accepted by the supplier (Q_L) and the customer demand/consumption (*D*). The delivery time (*DT*) is the time between the moment an order is placed by the customer process until the moment the ordered parts are delivered.

If the distance between supplier and customer is very short (e.g. within reach), it will not be necessary to include a transport system between them. In any case, there is a very important concept, which is known as the *Consumption during the Delivery Time* (C_{DT}). Thus, for a customer with a consumption/Demand (*D*), whose supplier has a delivery time (*DT*), the consumption during the delivery time is given by.

$$C_{DT} = D \times DT$$

Another very relevant concept is the so-called **Reordering Point** (*RoP*), which indicates the amount of *WIP* between the supplier and the customer, which, when reached, triggers the act of sending an order or request to the supplier (so that the latter supplies more parts). In a first (more simplistic) analysis, we can say that only when the *WIP* between the supplier and the customer is equal to the aforementioned *consumption during the delivery period*, a request or order must be sent to the supplier. In other words, in this first analysis, we are in fact considering that the reordering point is equal to the consumption during the delivery time, i.e.:

$$RoP = C_{DT} = D \times DT$$

As an example, let us imagine the following scenario:

- Customer Demand, $D = 30$ *parts/h* $= 0.5$ *parts/min*;

- Supplier delivery time, DT = 2 *min*;
- Supplier minimum batch, $Q_B = 1$ *part.*

The Reordering Point would be:

$$RoP = C_{DT} = 0.5 \times 2 = 1\ part$$

That is, an order must be sent to the supplier when the WIP between processes is 1 part. Here we would be in a production that could be on-piece-flow, since the supplier's delivery time (*DT = 2 min*) is equal to the customer's cycle time ($T_C = 1/0{,}5 = 2$ *min*). In this way, but assuming that there are no unforeseen events, it is guaranteed that the customer always has parts to work with.

Let us now consider a new scenario, still very predictable, but slightly different in terms of delivery time and supplier minimum batch:

- Customer demand, $D = 30$ *parts/h* = 0.5 *parts/min*;
- Supplier delivery time, DT = 20 *min*;
- Supplier minimum batch, $Q_B = 10$ *parts.*

In this case, one-piece-flow production would not be possible, since the supplier's delivery time (DT = 20 min) is greater than the customer's cycle time ($T_C = 1/0.5 = 2$ min) and the order point would then be:

$$RoP = C_{DT} = 0.5 \times 20 = 10\ part$$

In practical terms, we could recommend the so-called "Two-Bin System" (described in its own subsection below), in this case with 10 parts in each bin (Figure 4.9). In this way, starting with two bins full at the customer, the supplier stops supplying until the first bin is empty and is sent back to the supplier. At that moment, the bin with 10 parts remaining in the customer

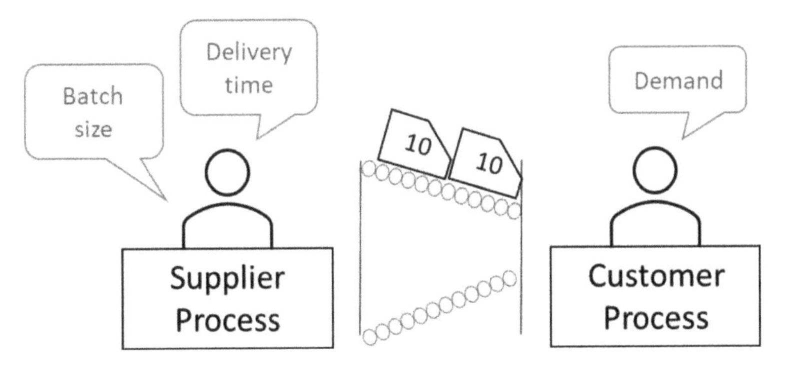

Figure 4.9 Two bin system.

buffer (or lineside rack or storage) is enough to meet the demand during the 20-minute delivery period (since it consumes one part every 2 minutes).

Now let us add some more realism (existence of variability and unpredictability) to the case assuming that the delivery time is not always the same and that the customer's demand can also vary. In these cases, there is a need to consider a Safety Stock (S_S) to deal with that uncertainty. Thus, the reordering point (RoP) is no longer equal to the consumption during the delivery time (DT) since it is necessary to add a safety stock (S_S) to it, ie:

$$RoT = C_{DT} + S_S = D \times DT + S_S$$

The safety stock should be as large as the variation in the supplier's delivery times and the customer's consumption. In-depth analysis requires data and statistical treatment, and it depends a lot on the type of risk the client is willing to take. Of course, it is not possible to protect a client process from everything that can happen in the supplier process, there is no level of WIP that guarantees such full protection (i.e. total avoidance of material shortage). As it is easy to imagine, there are events that can occur inside the supplier production unit or in the logistic system, which cannot be considered in the determination of the adequate safety stock level. It will not make much sense to safeguard stocks in case there is an earthquake in the area of the raw material supplier or even possible strikes by truck drivers. Basically, there is a certain allowable level of variation that we have to define when we want to determine the safety stock levels for any product, material or part.

We will then present two practical approaches for the determination of the safety stock values. To do so, we will assume, as an example, some hypothetical data concerning historical values of delivery time and demand presented in Table 4.2.

According to these historical data, the average demand (D_{avg}) and maximum demand (D_{max}) recorded were 30 and 40 part per hour, respectively, and the average delivery time (DT_{avg}) and maximum delivery time (DT_{max}) were 4 and 8 hours, respectively. Let us also assume that these maximum values for

Table 4.2 Historical data on demand and delivery times.

Week	1	2	3	4	5	Average	Max
Average demand (parts/hour)	28	31	28	32	31	30	
Maximum demand (parts/hour)	34	40	32	38	34		40
Average delivery time (hours)	4	3	6	3	4	4	
Maximum delivery time (hours)	8	6	8	5	7		8

demand and delivery times have variations within normal patterns (i.e., they are considered not to result from extraordinary causes). Basically, the manager assumes that the safety stock levels should respond to this type of variation. For this, two practical approaches can be adopted, which are as follows:

$$S_S = D_{max}\,(DT_{max} - DT_{avg})$$

Or,

$$S_S = (D_{max} \times DT_{max}) - (D_{avg} \times DT_{avg})$$

Thus, if the first approach is adopted, the safety stock would be:

$$S_S = 40 \times (8 - 4) = 160 \text{ parts.}$$

On the other hand if the second approach it adopted then the safety stock level would be:

$$S_S = 40 \times 8 - 30 \times 4 = 200 \text{ parts.}$$

The first calculation option tends to be a little more risky than the second and the reader will have to decide which of the two is more appropriate in his/her context. One possible approach might be to start with the second form of calculation and if stock levels remain perfectly controlled then move on to using the first equation.

If we assumed a safety stock of 160 parts, as suggested by the first SS calculation option, and the average values of consumption and delivery time, the reordering point would be:

$$RoP = D_{avg} \times DT_{avg} + S_S = 30 \times 4 + 160 = 280\ parts.$$

At these circumstances the customer process has to send an order to the supplier when its WIP level reaches 280 parts. Assuming for instance that the parts are transported in boxes with 100 parts each then the practical moment to send an order to the supplier would be when there is only 3 boxes in the buffer. An important question that must be put now so the reader ca go a bit further in the understanding of this mechanism, is: how many boxes should be ordered from the supplier? Does it work if the customer orders only one box at a time? To help answer these questions, look at Figure 4.10.

If we assume that, at a given instant, we have 4 boxes full of parts at the customer buffer and that reordering point is reached (3 full boxes left)

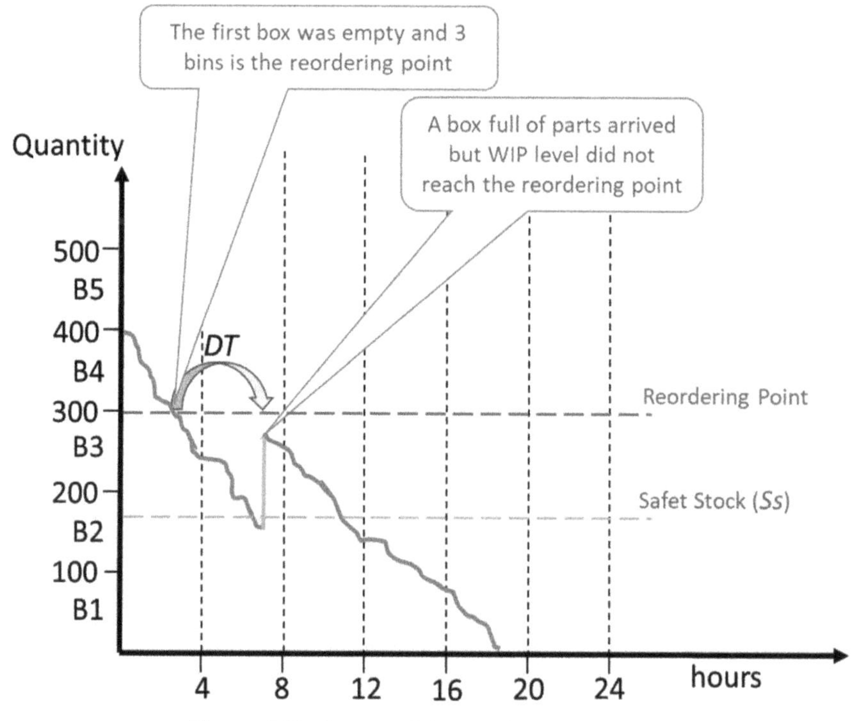

Figure 4.10 Impact of the quantity to be ordered.

as shown in Figure 4.10, sometime before the first 4 hours have passed. At that moment an order is placed so the supplier takes 4 hours to respond to the request. Four hours later a box full of parts arrives when parts are already being consumed from box B2. At that moment, even with that just arriving the box, the total stock at that moment is lower than the order point; therefore, no new order is placed and the system collapses (lack of parts). To solve this problem, the quantity to be ordered from the supplier should be increased to 3 boxes (equal to the order point). In this way, even if consumption is maximum and the delivery time is not met, when the delivery is made the stock level will be above the order point.

In this case it would make sense to use a *kanban* system[5], assigning to each box a (*kanban*) card. A *Kanban* system would perform better than the reordering point in terms of average stock levels since a request is sent to the supplier every time a box become empty.

[5] More information on Kanban systems is given in the annex.

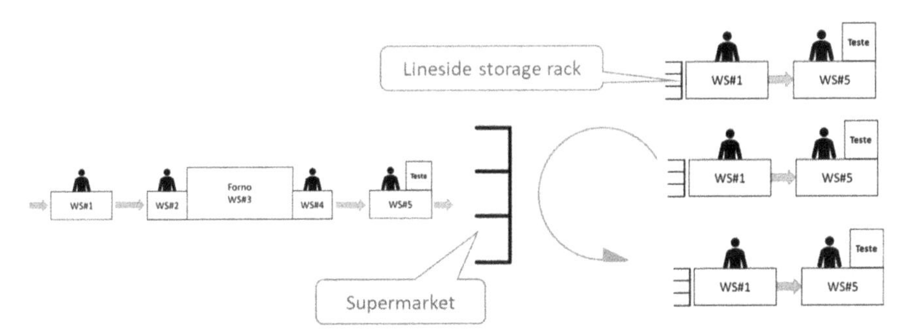

Figure 4.11 Supermarket between a supplier process and several customer processes.

In general terms, for boxes or bins with quantity capacity, , and for a consumption, D, delivery time, DT, and safety stock, , the number of *kanbans* required is:

$$n_K = \frac{D \times DT + S_S}{Q_{Bx}}$$

In our previous case we have = 280/100 = 2.8 *kanbans*, which in reality is 3 *kanbans*. There is an assumption to keep in mind when calculating the number of *kanbans* to circulate between a supplier and a customer: when the customer starts consuming parts from a box, the worker must immediately send the Kanban back to the supplier. If the *kanban* card is glued to the box, and since the box will only be sent to the supplier process when it is empty, then it will be necessary to add a box to the circuit and obviously glue a *kanban* card to it (4 boxes, 4 cards).

Now, let us assume a more complex case where the supplier supplies several types of parts to several customers. In this case, continuing to ensure pull production, there is a popular good solution called "supermarket"[6], as illustrated in Figure 4.11.

Supermarkets are different from traditional warehouses in many ways. Supermarkets must store boxes preferably on "dynamic" shelves, i.e. which take advantage of the force of gravity, and also guarantee the following:

- Single (rack) position for each part reference;
- FIFO discipline;
- Replacement triggered by consumption;

[6] More information about supermarkets is provided in the annex.

- Easy access to boxes (containing parts/materials/ components/ sub-assemblies);
- Easy visualization using visual management.

Each supermarket shelf (dedicated to a part/material/... reference) must have enough space to store the maximum quantity that can, in extreme cases, exist in the supermarket. This maximum quantity obtained by adding the order point of this reference to the quantity that is ordered from the supplier, i.e:

$$Q_{Max} = RoP + Q_O$$

The quantity that must be ordered from the supplier, here called order quantity (Q_O) is here understood as the quantity that must be ordered from the supplier whenever the stock quantity drops to the reordering point. This quantity should be such that it guarantees that there will be no material shortage, while at the same time ensuring that the maximum quantity in the supermarket is as small as possible. The order quantity is then determined as the maximum value between the supplier's minimum quantity to be delivered (Q_D) and the order point itself (*RoP*), i.e:

$$Q_O = \max(Q_{D,RoP})$$

The maximum possible quantity in the supermarket for a given reference should include the reordering point (*RoP*) for that reference as well as the respective order quantity (Q_O). That maximum quantity is given by the following equation:

$$Q_{Max} = RoP + Q_O = RoP + max(Q_{O,RoP})$$

Based on this maximum quantity, the length of the supermarket rack for a given type of part is determined by the maximum quantity of boxes and the size of each box (see Figure 4.12).

4.6 Two-Bin System

The two-bin systems are very simple systems that guarantee good efficiency in the management of material flows, and for this reason it has been deemed appropriate to present it here. These systems can also be used very effectively outside the industrial context, for example in hospital materials management, office materials management and many other contexts. The example presented in Figure 4.13 shows a two-bin system implemented in the department where we work (Production and Systems Department of the Engineering

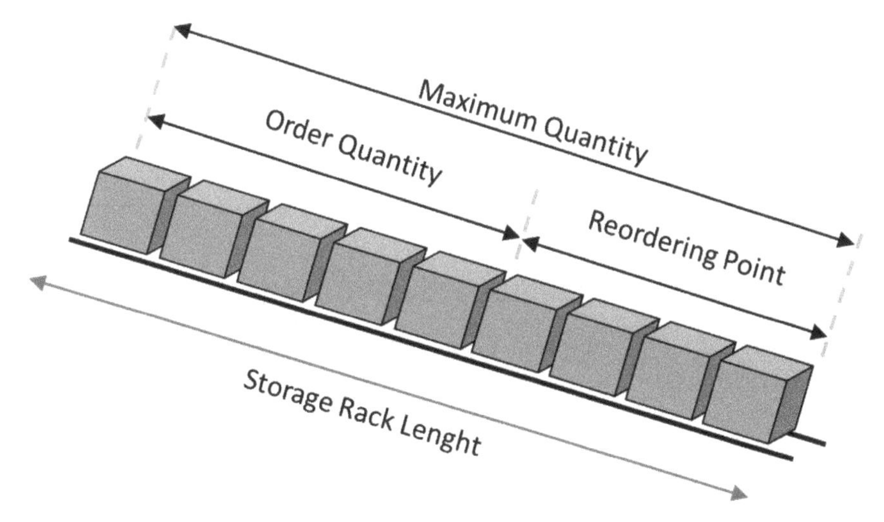

Figure 4.12 Supermarket sizing.

Figure 4.13 Two-bin system in office area in our department.

School of the University of Minho) and has greatly simplified the work of the person responsible for managing these materials (ball-point pens, post-its, glues, paper clips, markers, etc.).

Our department's staff, both teaching and non-teaching personnel, stock up on these materials according to their needs (consumption) and, when someone removes the last unit from a box, they only have to place that (empty) box on the top right shelf, marked with "*Abastecer*" (meaning "Supply"). In this particular case, every Monday, the person in charge of this material management replenishes the empty boxes, placing them back (full) in their respective positions. Each full box that goes back to the shelves is placed behind the respective box that is already there being consumed.

4.7 Obstacles to One-Piece-Flow Production

There is no doubt that one-piece-flow production is desirable and should be pursued continuously by continuous improvement systems. It is however important to recognize that there are always various types of obstacles in the way that make its achievement very difficult. A first obstacle that needs to be tackled is the frequent difficulty that employees and supervisors have in understanding the virtues and advantages of this paradigm. This first barrier is not easy to break and there are cases that require some art and persistence to take steps towards its successful implementation. It is important to mention that employees should not be forced to accept such a drastic paradigm change, moving from traditional production (with storage of products waiting between processes) to a paradigm of production in a tense and continuous flow (without "buffering" between processes). A very interesting case occurred in an assembly cell where we wanted to implement one-piece-flow production and that deserves to be mentioned here.

In an assembly cell where a team of 4 employees normally worked, we tried to show them (in some training sessions and conversations in the production area) the virtues and advantages of one-piece-flow production in terms of team performance. We tried to show that this type of production would allow us to respond much more quickly to orders, but the employees were never really interested in adopting this way of working. We never wanted to force its implementation because we understood that the will to experiment should come from them. Anyway, we accompanied the team in a continuous improvement effort, identifying and implementing improvement opportunities, using an adapted Toyota Kata[7] approach. After some improvements in the production cell layout, new formats and layout of the racks for storage of products in progress (WIP) and a system to control the product flows, we achieved

[7] Toyota Kata approach will be covered in chapter 7.

some interesting gains in the reduction of the throughput times of that cell and, consequently, an increase in the response speed. With these improvements, we managed to go from an average throughput time of around 6 hours to a little over 2 hours (i.e., the cell started to respond much more quickly to orders). These gains were achieved over about 6 months, with an increasingly ambitious goal being set as the previous goal was reached. When we reached just over 2 hours (already a very good result), the team realized that they could not reduce the throughput time even further without radically changing the way they were working. At that time, they decided to implement the so-called "Rabbit Chase" operative mode as a way to work in the cell. In this way of working, each operator performs all operations on each product. After performing the first set of operations at the first workstation, the employee takes the product to the next workstation where he performs the respective operations, and then moving to the third workstation and so on until the last workstation where he completes the product. Then the operator returns to the first workstation to start the production of next product. In this way, the team arrived at the implementation of one-piece-flow production in a natural way. The throughput time was reduced to value under one hour with this approach.

The other type of obstacles that often hinder the adoption of one-piece-flow production is a little more justifiable and is related to the nature of the processes themselves. An example of this is the relationship between the distance between two consecutive processes and the cycle time. The greater the distance between two processes, the greater the restriction caused by the quantity to be transported each time. If we have, for example, 20 meters between two consecutive processes on parts that are produced at a rate of 60 per hour, it will be difficult to transport one part at a time. In this case, it is necessary to accumulating some parts and then transport them to the next process. Of course we should find solutions to bring these two processes closer together, there is no doubt about that, but there are many cases that make this closer together difficult or impossible. It is important to note that if the cycle time was one day, then that same distance of 20 meters would no longer be a problem and it would be perfectly feasible to transport one piece at a time.

Another obstacle to one-piece-flow production is the time required to perform the setup of the equipment. There are processes that, due to their technological nature, do not allow producing only one piece at a time at acceptable costs. If we consider, for example, a plastic injection machine with a cycle time of 20 seconds per part, it would be unfeasible to inject just one part, change the injection mold, produce just one unit of another type of part, change the injection mold again and so on. In processes involving injection machines, metal stamping presses, or others of the same type, the technology that currently exists makes one-piece-flow production unviable for most products. It is, however, imperative that solutions are sought to reduce setup times as far as possible so that it is feasible to produce smaller quantities at a time and thereby increase fluidity. The classic methodology for reducing setup time is called SMED (*Single Minute Exchange of Die*)[8]. Its importance in the TPS / Lean context is such that the person that is frequently known as its creation, Shigeo Shingo, after having spent 15 years developing it, dedicated a book to it with the title "A Revolution in Manufacturing: The SMED System" (Shingo, 1985). Automation and robotics have also been helping to reduce setup times and, in addition, new technologies, such as 3D printing or others that may be developed, may change the production paradigm so that there is no advantage in producing in batches.

To conclude, another type of obstacle to fluidity and one-piece-flow production has to do with the nature of the technology of some processes that clearly forces batch production. An example of this is the cutting of fabric, textile canvas, metal sheets or paper. Productivity in cutting is achieved by cutting several layers of the same material at once. The cutting time is the same, whether with one layer or several. In this way, it is usually cheaper to cut a set of identical sheets and therefore batch production becomes more attractive.

4.8 Concept of Flow in Domestic and Personal Life

Although not as visible in our domestic and personal life, the concept of flow also applies effectively. The stagnation of things waiting to be dealt with is, just like in an industrial environment, the opposite of flow. Will the reader be familiar with typical cases like, the tap in the bathroom needs to be fixed or the garage is all messed up? And how we put these tasks off as long as possible? One of the problems is that it affects us continuously, it's something that we carry around with us and that disturbs/ bothers us because, from time to time, it appears in our mind reminding us. As if that wasn't enough, there is

[8] This technique will be briefly presented in the Annex.

always someone, from our family or friends who will make a nasty comment about it to remind us, right? And we always have the old excuse that we don't have time.

The idea of the concept of flow in our domestic and personal life lies in its great practical advantage. Whenever we finally finish one of those tasks we should have already done, we get a huge sense of relief and comfort. It is as if we have refilled ourselves with energy or as if we have freed up an occupied space in our mind.

Instead of waiting for the day (or morning or afternoon) when we will eventually have time to dedicate to one of these delayed tasks, knowing that this opportunity may take a long time to occur, it is better to start taking small steps with some frequency. It's equivalent to the concept of batch versus one-piece-flow production. Making small increments every day (or almost) on those tasks that are left undone for a long time has something similar to one-piece-flow production. The feeling is as if we are in control of the process rather than the impression of having a burden that is left by the tasks we have overdue. Admittedly this is not always possible, there are cases where it is not possible to divide a big task into small tasks. If you have to paint a room at home, you'll have to dedicate a relevant amount of time to it, like a morning or an afternoon. But in cases where it is possible, the concept of fluidity is a concept that also works effectively in our personal life. Let's use the case of the garage, the office or the messy desk as an example. If every day we remove or tidy up just one item (a document, a tool, a pen) that is too much or out of its place, a month later the difference will be enormous. Without great effort, just by making small increments, extraordinary things can be achieved. This is the concept of fluidity in our domestic and personal life. You just have to experiment and practice it.

References

Alukal, G., & Manos, A. (2006). *Lean Kaizen: A Simplified Approach to Process Improvements*. Milwaukee: ASQ Quality Press.

Beecroft, G. D., Duffy, G. L., & Moran, J. W. (2003). *The Executive Guide to Improvement and Change*. Qualilty Press.

Jones, D., & Womack, J. (2002). *Seeing the Whole. Lean Enterprise Institute, Brookline*. Cambridge, MA, USA: Lean Enterprises Inst Inc.

Little, J. D. C. (1961). A Proof for the Queuing Formula: $L = \lambda W$. *Operations Research*, *9*(3). https://doi.org/10.1287/opre.9.3.383.

Markovitz, D. (2011). *A Factory of One: Applying Lean Principles to Banish Waste and Improve Your Personal Performance*. Productivity Press.

PPDT, P. P. D. T. (1998). *Just-in-Time for Operators (The Shopfloor Series).* Productivity Press.

Rother, M., & Shook, J. (1999). *Learning to see: Value stream mapping to add value and eliminate muda. The Lean Enterprise Institute.* https://doi.org/10.1109/6.490058.

Shingo, S. (1985). *A Revolution in Manufacturing: The SMED System.* Oregon: Productivity Press.

5

Decoding Continuous Improvement

Continuous Improvement is a powerful concept; however, it is very often misinterpreted and poorly implemented. This chapter aims to clarify as much as possible, what Continuous Improvement is and which are the main mistakes in its interpretation. For Continuous Improvement to become a reality there is a lot to do or to change in organizations. Continuous Improvement must be perfectly integrated in the daily routines of the organizations in order to be efficient and effective. As such, it is a powerful tool to increase the competitiveness and sustainability of organizations. For that to happen, the vision of the organization assumed by top management and shared by all, must explicitly include Continuous Improvement and an entire system of routines must exist or be created.

5.1 Hoshin Kanri

Before we move directly to the intricacies of Continuous Improvement we must remember that it only makes sense when the direction in which we want the improvements to occur is known. Only after having defined the direction or "True North", as Figure 5.1 intends to represent, it will be effective to start implementing continuous improvement in the organization or company. Besides defining this direction, it is also necessary to share and decode it throughout the organization according to the local language of each area. *Hoshin Kanri* is a mechanism that helps to define in a detailed way a set of strategic decisions, the planning of actions and their control/management, in order to keep the whole organization aligned with the vision of its main decision-makers.

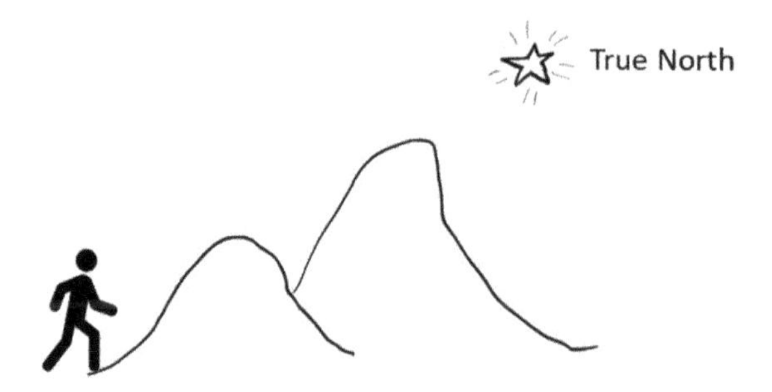

Figure 5.1 Direction, vision, "true north".

Hoshin Kanri is a strategic planning and management methodology centered on the concept of PDCA cycles (*Plan-Do-Check-Act*). Its authorship is attributed to Professor Yoji Akao, a Japanese specialist in planning, who published the book "*Hoshin Kanri: Policy Deployment for Successful TQM*" (Akao, 1990). In this book, the *Hoshin Kanri* methodology was applied in a context where the strategy was very much a quality statement, but it can in fact be used in any strategic endeavor. The Japanese word *Hoshin* means, in this context, "direction" and the word *Kanri* means "control" or "management". This methodology is used to create goals, assign intermediate measurable milestones to those goals, and, monitor and evaluate their progress over time.

This technique is used by Toyota to materialize strategy and helps strategic alignment throughout the organization. *Hoshin Kanri* is much more than the well-known strategic deployment (*Policy Deployment*) to which it is too often given equivalence. In addition to this deployment function, *Hoshin Kanri* also includes strategic vision, strategic planning and the mechanisms for controlling or managing that strategy. Strategic planning for a given period is the description of what should be achieved, and also how and when, in alignment with the vision. The strategic control or management is how the execution of the plan will be monitored and managed.

Strategic deployment means passing on the vision and strategy defined by top management to all divisions, departments, sections and operational teams. On the one hand, we have the development of the strategy and on the other, we must also get the organization, in its day-to-day operation, to do what is necessary for that strategy to be followed and achieve the objectives set out. This tool may also be described as the way to ensure that the organization's strategy is executed throughout its hierarchical structure. This does not always mean that it is enough to communicate the strategy to all levels of the organization, from top management to operations. Although

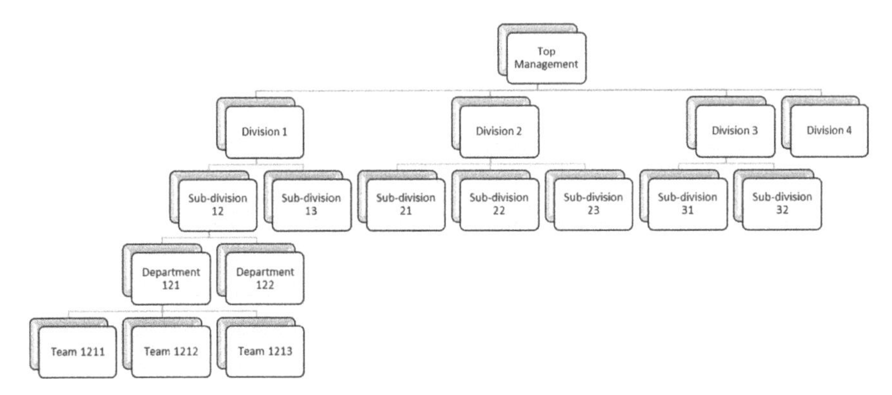

Figure 5.2 Strategic deployment.

each organization finds its own way of unfolding strategy in its structure, as Figure 5.2 attempts to represent, there will be two quite different approaches.

One approach implies that each entity (division, department, team) adapts to its context the strategic plans and objectives of the entity at the hierarchical level immediately above. Using the hierarchical representation in Figure 5.2 as a reference, Division 1, knowing the strategic objectives of the Top Management, will create its own strategic objectives and indicators, adapted to its reality. In turn, Sub-division 12, knowing the strategic objectives of Division 1 will assume the same procedure. Regarding the alignment let's see an example: The Department 121 established as one of the objectives the reduction of the lead time in its value chain and Team 1211 in alignment with the department to which it belongs (Department 121), established as an objective the reduction of the tool change time (setup) in its main equipment (presses for example). Although the indicators are not the same, the reduction of setup time contributes to the reduction of the department's lead time.

Another possible approach, successfully used by some large multinational organizations, is to develop an overall strategy for the organization, then communicate it throughout the structure and let each entity in the hierarchical structure create its own goals aligned with the overall strategy. In this approach, each unit does not have to align with the strategic goals of the hierarchical level immediately above but only take inspiration from the overall organization strategy.

As an example, Figure 5.3 shows the *Hoshin Kanri* of the Lexus factory in 2015, provided by a Toyota employee[1]. The idea is by no means to divulge Toyota's secrets, especially since it is a document already 5 years old, but

[1] This person prefers not to be referred.

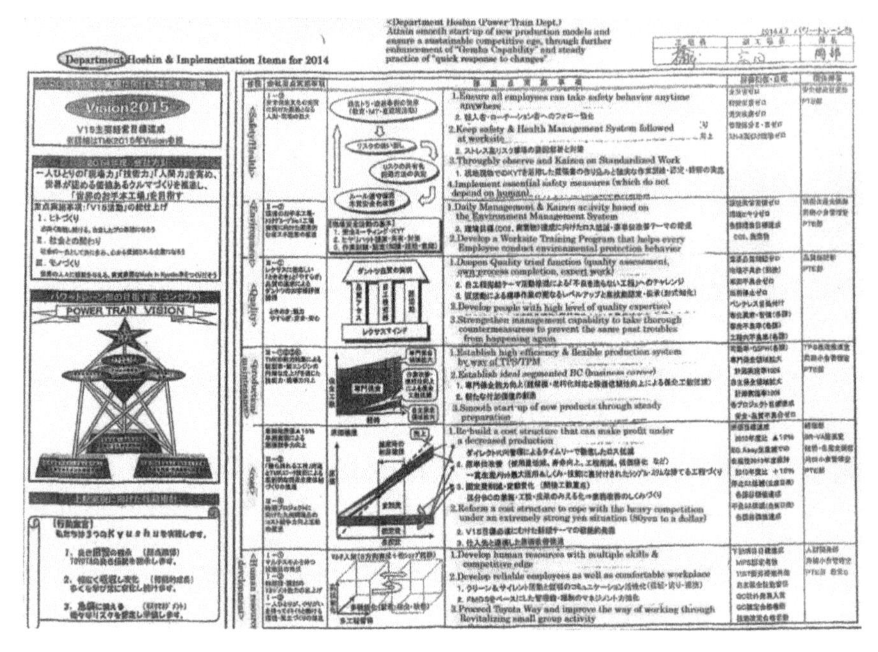

Figure 5.3 Hoshin Kanri of Lexus factory in 2015.

only to show evidence of Toyota's *Hoshin Kanri* practice and the logic of its use. It is curious to see the importance of the visual management achieved with the diagrams and graphics that appear in the document. Toyota's constant quest for clarity in communication, using pictures instead of text whenever possible, is also reflected in this document.

This way of organizing and standardizing strategic planning and management includes the following elements: Vision, goals for the coming year, detailed action plan and monitoring mechanisms. To understand better this tool it might help to present an example of a personal Hoshin Kanri from the author. Yes, this methodology can also be used in our personal life.

As you can see, Table 5.1 presents an example of *Hoshin Kanri* for one of the authors of this book. In the first column, there is a retrospection on what happened the previous year and all the other columns are concerned with planning for the year 2021. The major areas are defined, which in this case are health, projects and teaching and then the objectives and goals are detailed as well as the actions that will be necessary to achieve them and how they should be controlled. This personal "strategic planning" process is extremely important because it makes us reflect on the path we want our lives to take and allows us to invest in what is important. Throughout the year, we

Table 5.1 Personal Hoshin Kanri 2021

2021 Vision:

- *Health: Maintain physical and mental health. Watch out for food and drink, keep exercising, work fewer hours, train patience and try to be more open-minded.*
- *Projects: To write a book on continuous improvement in English. Apply for more funded projects to progress in research. Maintain projects with Inditex.*
- *Teaching: Improve the quality of lessons and student support and improve interaction with other colleagues in the department.*

	2020 Balance	2021 Vision	2021 Goals	2021 Action plan	Frequency
Health	• Some mistakes in eating and drinking - I need to correct in 2021 • Overloaded periods of work • I have not always kept up regular physical exercise	• Maintaining good health ○ Controlling overeating ○ Controlling blood pressure ○ Keep up morale	• Keep your blood pressure under control • Run and walk 5 km a day • Eat early and go to bed early	• Check healthy eating • Check that I run or walk 5km a day in rain or shine • Check that I have been patient and open-minded	• Check daily
Projects	• Growth in Lean Hospitals • Growth in Operational Excellence	• Increase contribution to the hospital • Invest in funded projects • Publish scientific articles	• Achieve significant improvements in the hospital • Submit 3 applications • Submit 5 journal articles	• Implement actions every week • Search for programs by March. • rite an article every two months	• Check progress every week
	• Book on continuous improvement ready • A project success story with Inditex	• Publish a book on continuous improvement. • Write book on CI in English • Augment success stories	• Book on sale by March • Write book in English by June • Obtain 2 success stories	• Check progress • Search and select publisher for publication in English • Select only potential cases of success	• Weekly check • Weekly control • Monthly control
Teaching	• Entry into projects with social organizations • Insufficient performance in the quality of the relationship with students	• Growing in service learning • Improve relationships with students	• Implement ApS projects in 5 social organizations • Achieve > 4 in surveys (all CUs)	• Achieve 5 ApS projects in the SPL LU and 6 in the EL LU • Improve quality of OSPii classes • Maintain quality in other CUs	• Check in March • Monthly check • Monthly check

are constantly asked for unimportant but urgent tasks that rob us of much of our time and hold us back from the truly important tasks.

The strategic management part is materialized in the control of the execution of the planned actions according to the last column. This process is closely linked to the PDCA cycles used in different time dimensions. We can easily identify an annual PDCA cycle intrinsic to the *Hoshin Kanri* process itself. A plan is made for the next year, then this plan is executed and at the end of the year, a retrospective has to be made (first column of Table 5.1). With the learning of what went well as well as what went less well, make some decisions about the new annual cycle. On the other hand, there will always be small PDCA cycles related to the control frequency defined in the last column of the same table. It will be good practice for each of the daily, weekly or monthly controls to have natural PDCA cycles. This way it will be easier to guarantee that the plan will be fulfilled. However, the plan may be reformulated once or twice during the year if there are unexpected disturbances.

The personal *Hoshin Kanri* is already an excellent tool to help plan and manage our personal life, but, in organizations, its effect is much greater. In organizations, two characteristics boost the effectiveness of this methodology: firstly, the fact that there may be teams and secondly, the fact that the strategy may be deployed at various levels of the organization. In organizations, teams do strategic planning, execution and control with the typical gain created by their synergies. It is not just one person thinking and there may be one person who is less motivated at certain times but there is always another who compensates. When we have a team, it is easier for someone to worry about control and collection in the fulfilment of the plans and to seek some action in relation to possible deviations. In personal *Hoshin Kanri* the execution and control is carried out by the same person with the inherent difficulties. It takes a lot of discipline to make this methodology work effectively, as the same person is responsible for planning, execution, control and accountability. The other possible advantage in organizations is the fact that there is another element, which is the strategic unfolding or passing of strategic information throughout the hierarchical structure. This allows each of the entities (departments, sections, operational teams) in the face of the *Hoshin Kanri* of the entity immediately above to create more easily their own in an effective way. There is certainly a contribution from each to the strategic objectives of the others.

Additionally, the idea of strategic alignment, which is achieved with *Hoshin Kanri* and strategic deployment, is to create a platform and a framework in which all elements of the organization share aligned objectives, have a vision of the whole, feel that they are part of it and feel that their contribution is fundamental to everyone's objective. This framework enables and motivates all employees, with their commitment and contribution of ideas, to improve their workplaces, their sections and departments and brings more than just innovation and improvement to organizations. This framework gives employees a sense of importance in the workplace and importance in the whole when their suggestions are put into practice.

Here is an example of informal strategic misalignment. In a organization whose name we will not disclose we came across a team that reduced the speed of an important piece of equipment in its production unit, whenever they felt that their target production would be reached without the need to resort to overtime. Since their goal was to work overtime to earn more money, they were contradicting one of the organization's goals, which was to increase the productivity of that production unit.

An example of strategic alignment, that can be given here, concerns the multinational organization here called Company_B that clearly defined that one of their main objectives, for their factory in Guimarães for the year 2019, was to reduce accidents and incidents. As they consider people first and as some of the processes present, by their nature, some risk of accidents they decided that safety would be the most precious indicator to improve. In this sense, all sections and departments monitor and implement actions so that this indicator improves over time. Likewise, the operational teams of each of these departments and sections have also internalized this indicator, constantly monitoring it and implementing actions so that it is reduced. In all daily meetings of the operational teams, the importance of safety and the need to improve this indicator are mentioned. In this way, the whole organization keeps its focus on safety. Other indicators are also monitored in the background such as quality indicators, then productivity and cost indicators. When they are relatively satisfied with the achieved reduction of accidents and incidents then they can change the main goal.

Other organizations, such as the multinational organization here called Company_C, shift their focus to continuous improvement every six months and deploy this new strategic focus in their operational teams. In one six-month period, they may be focused on quality because a gap has been identified in that area and in the next period they may be focused on productivity.

5.1.1 Definition of improvement

The action of implementing a change with the intention of translating it into an improvement does not always effectively result in an improvement. It happens that sometimes the implemented change does not result as expected in an improvement but in a failure. Furthermore, there is a subjective part in the judgement since not everyone accepts that a certain change is in fact an improvement. Some may think it is an improvement but others may have a contrary opinion. This kind of conflict is constant in our day-to-day life and therefore we cannot lightly assume that an improvement we want to implement in a process or in a production unit is an improvement on all points of view. It is very natural that in traditional organizations there are often arguments about whether a proposed action or a proposed change will lead to a result considered as an improvement. Some implemented improvement actions can bring advantages on one side but it is also natural that they bring disadvantages on the other side. The intention is that the advantages are greater than the disadvantages. Furthermore, even if a change brings apparent advantages for all, it may not work, simply because there is a negative predisposition on the part of the people who are the target of this "improvement". In organizations where there are some latent conflicts between top management and employees, there may be this kind of negative predisposition towards any proposed change in the way of working. In this kind of environment, it is usually very difficult to achieve changes in behavior and to make changes even if it is clear that they will result in improvements for everybody.

Going back to the question regarding situations where a change is perceived by some as an improvement but not by others, let us see an example: consider an improvement action that leads to an increase in productivity and a reduction in lead time, but its implementation requires an increase in the multi-skilling of employees, in a organization with a very strong culture of work specialization. This type of conflict shows the misalignment and lack of clear vision of an organization. The increase in the diversity of employees' skills is sometimes perceived by many as an obstacle or a threat. There will be many more examples, but the message is that improvement is only improvement if it happens, as mentioned before, in the direction that makes sense for each one. Hence, it is necessary to clarify very well which path each organization wants to follow, where it wants to go, in short what its vision is.

It is vital to know where the organization wants to go, for example, which key indicators (KPI - *Key Performance Indicators*) the organization wants to improve and is committed to doing so. There are cases in which the person responsible for deciding on an improvement proposal, without

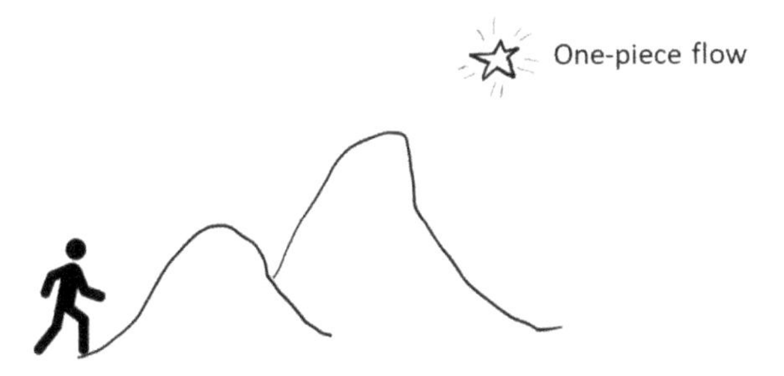

Figure 5.4 The long-term goal is the one-piece flow.

knowing it properly, makes comments like: "yes, yes, let's do it, we are always willing to improve" or "anything to improve is welcome". These and similar ways of approaching the issue, although frequent and promising, are not much help to those who want to make improvements. This can demonstrate disinterest with the vision and purpose of improvement. It is important to judge any proposed improvement from the point of view of its alignment with the vision shared by all. Furthermore, it should be borne in mind that a proposal for change only becomes a real improvement if it is implemented and accepted by the people who will have to live with that change.

When the direction to follow is known and shared by all, it is easier to choose the type of improvement proposals to be accepted for implementation. If the organization wants to improve fluidity and increase the speed of response to the market, then everyone will understand any proposal that improves that fluidity as an improvement. However, it may conclude that such a proposal is too expensive to implement and so it is decided that it will not be implemented at this time.

Let us imagine an organization that has set itself the goal of constantly increasing fluidity and reducing WIP until it reaches, in the distant future, one-piece flow production (Figure 5.4). In this organization, the purchase of a new piece of equipment that results in lower cost per piece but increases the lot size should not be perceived as an improvement. That organization will more quickly be able to buy equipment that will reduce the lot size even if the cost per piece in that operation is higher. It may seem strange but the gains from increased fluidity bring overall cost reduction in the medium and long term. Here it makes sense to remember the first principle of the Toyota Way, which says, "Base your management decisions on a long-term philosophy, even at the expense of short-term financial goals".

As an absurd example, let us suppose that an organization has as its strategic goal to improve the use of all its equipment - that is its vision. It is in this rather classic way of thinking that the organization thinks it can lower costs and be more competitive and sustainable (it does not want to have idle equipment because it wants to make their use profitable). According to this objective, all actions that result in increased equipment usage are interpreted as improvements. However, the truth is that these "improvements" may be increasing stocks, worsening quality, increasing response times and therefore reducing competitiveness. That is why improvement has to be guided by a direction (the "true north"), vision, purpose and values, which are appropriate for the long-term survival and sustainability of the organization. Another example that illustrates the mistakes that can be made about what can or cannot be an improvement is the following: let us imagine that a change introduced in an assembly line resulted in a productivity increase. Now imagine that this change also degraded ergonomic aspects or increased the risk of accidents. Can this still be considered an improvement?

Regarding failed attempts at improvement, we have to accept that some changes that someone proposes to make and that apparently bring advantages and improvements, may end up not working. This happens because some variables were not, or could not be, taken into consideration when the proposed improvement was accepted. Some of the ideas that seem to work, sometimes simply do not work. Everyone should accept this fact with serenity in a positive way because it always results in valuable learning for the team and the organization. From that failure sometimes better ideas emerge that would not have been generated if you had not gone through that experience. The openness to accept that some improvement attempts will fail is important so that everyone feels free to propose.

5.1.2 Continuous improvement *versus* occasional improvements

When people start talking about implementing continuous improvement in an organization comments are often made with apparent assurance such as "... but we are always improving" or "we have already made many improvements over time". In addition, they give examples of what it was like before and how they have evolved. "We used to function badly and full of problems but now we function much better". There are also unfortunately those who say that it is very difficult to improve because they have already made all the improvements that were needed; but that is another matter.

The argument that they have already made many improvements and are always improving can be for some the answer to the question: "why do we need continuous improvement if we have already been doing it"? The truth is that quite naturally organizations will make improvements over time. However, are they improving performance at a faster rate than their competitors are? Or at a speed that will guarantee them greater gains in the future? The main idea behind implementing a continuous improvement system is to adopt routines and methods that ensure that improvements are created and implemented in a consistent, sustainable and stable way over time.

There are many examples we know of where key decision-makers refuse to accept that the disruptive improvements they achieved in the past no longer guarantee competitiveness today. We can mention here two cases that are essentially the same. The managers of an important SME in the center of the country and the manager of an SME from another sector of activity in the north that we visited a few years ago asked us for help because they felt that they were losing competitiveness with each passing year. Interestingly, they refused to accept that the way they organized and managed their production was outdated. They were using a technology that gave them a huge productivity gain in the 1980s and because they thought it was optimized they couldn't accept that that disruptive improvement, which took place so many years ago, had lost its effectiveness in the current market conditions. Our proposals were not accepted because they refused to accept that the technology they had was no longer competitive.

In the same line of reasoning, it makes perfect sense to clarify the difference between doing one improvement or another and doing Continuous Improvement. Occasional disruptive improvements do not really create an effective culture of Continuous Improvement although many organizations still believe that making disruptive improvements here and there is the way to go. Getting an organization to have Continuous Improvement in its entirety is no easy task at all. Very large structural changes are required at the same time as changes in attitude, behavior, and as a result, culture.

Mike Rother in his book "*Toyota Kata*" Rother (2010), also agrees that workshops and improvement projects (we can also include the popular "Kaizen Events"), are not the same as Continuous Improvement. At Toyota, which is considered the organization that best carries out Continuous Improvement, the improvement process occurs daily in all processes and at all levels of the organization. Moreover, even when performance targets have already been met, the improvement effort is maintained.

Furthermore, Continuous Improvement can no longer be seen as just a factor of competitive advantage against the competition, but rather as a factor of survival. Organizations that do not continuously improve their performance will certainly succumb to their competitors. Continuous Improvement is no longer an option and has become the way to survive; the question is: how can each organization materialize Continuous Improvement?

Continuous Improvement, in its fullness, happens when everyone, every day, contributes with small improvements in their areas of influence. It is more desirable that small improvements are constantly implemented than occasionally implemented large improvements. When everyone continuously contributes to improve his or her areas in a way that is aligned with the larger strategy and purpose of the organization, everyone benefits from learning, everyone feels included and everyone grows with the organization. If the improvements were the result of large projects carried out by technicians and managers, there will hardly be the feeling of belonging and learning.

The difficulty to implement and maintain Continuous Improvement is enormous and few organizations manage to do it effectively. It is impossible to believe that continuous improvement can be achieved and maintained without effort and constant dedication and attention. It is a bit like maintaining equipment or maintaining relationships. It is necessary to maintain constant attention, dedication and care. This difficulty is reported, for example, in the book "*Toyota Way to Continuous Improvement*" by Jeffrey Liker and James Franz (Liker & Franz, 2011). In this book, the authors state that even Toyota has difficulties in maintaining Continuous Improvement, especially in Toyota plants outside Japan.

A possible first step towards Continuous Improvement may be to hire consultants to implement solutions that improve the performance of processes and flows. This path is viable in the first improvement interventions, but may not be viable in the medium term because, although in the first interventions the consultants can achieve with little effort significant performance gains, in the following actions they will achieve fewer and fewer gains. At a certain moment the cost inherent to consultancy may be higher than the small gains obtained and therefore this approach becomes unviable. Anyway, hiring consultants specialized in implementing improvements according to Lean, Kaizen, Toyota Way or any other model that seeks operational excellence, is very often the approach organizations choose to start adopting Continuous Improvement practices.

The next step is for many organizations the internal development of knowledge about the main Toyota-inspired excellence models like Lean philosophy, the Shingo Model and the Toyota Way, and in particular about continuous improvement. This is achieved by training a small group of managers in these areas and then giving them the role of drivers of improvement actions, with improvement events or other practices. As the organization becomes more mature in terms of Continuous Improvement, the number of managers and employees that dedicate part of their time to this type of activities increases more and more, until, in an ideal state, they all become drivers of Continuous Improvement. The trend is that more and more companies and other organizations start to include Continuous Improvement activities in the daily routines of all their employees. Organizations that do not do so will gradually lose market share.

5.1.3 Continuous improvement *versus* excellence

Before really getting into the subject, and in order to try to clarify what is going on in the world of designations linked to this area, let us try an analogy. In several types of products, there is the use of a brand to replace the name of the product. Many people say things like "I'm going to buy a Gillette", "I like your Jacuzzi", "do you want a Chiclet", "pass me the post-its", and many others. In these cases, we relax the language and collectively choose to use the brand name to designate a generic product. This happens even though there is a generic product name that can be used in most cases such as "razor" or "chewing gum". However, in the case of "Post-it" it may not be easy to find a generic product name that could be shared by most people. This is exactly the case with the generic product for the brands "Lean Thinking", "Kaizen", "Shingo Model", "Lean Six-Sigma", "World Class Manufacturing", "Quick Response Manufacturing", "Agile Manufacturing", and others. This is our understanding, but there may be other different ways of looking at this phenomenon. The challenge is to find the name of the "product" for which there are all these alternative brands. It is also acceptable that each of the authors of these brands have an interest in their brand name becoming universally accepted as the product name. The truth is that so far there does not seem to be a consensus on the generic name that brings all these alternative brands together. The term "Lean" may even be the term that has gathered the most popularity and for many people that term is already the product name but this is just one possible interpretation. At this moment we would be inclined to temporarily use the term "Excellence in Organizations" as the product name. So, to conclude this dilemma, in our interpretation both "Lean Thinking", "Toyota Way" "Kaizen", and "Shingo Model" are Toyota Inspired Excellence Models (TIEM).

To make things even more complicated, we often confuse Continuous Improvement with Lean Philosophy or with any other alternative of those

TIEM. This subject can be approached from several perspectives and may raise numerous discussions, but Continuous Improvement is a principle or concept that is included in the excellence models we have been talking about. If we use the Lean philosophy as a reference, in one of the publications of the research group responsible for the very name Lean (Womack & Jones, 1996), Continuous Improvement is only one of the five principles of the Lean way of thinking (Lean Thinking). Let us say that, the other four principles point the direction to which we should align the improvement efforts, that is, they indicate us where the Continuous Improvement should be directed. This subject of the principles of each one of the models was already presented in some detail in chapter 3.

Based on current knowledge, we believe that the most effective and successful direction, which should be used by organizations to align their Continuous Improvement work, is the one that is translated into the principles that underpin the main Toyota-inspired excellence models presented in chapter 3. A necessary factor for success occurs when the top decision-makers of an organization believe deeply that these principles are the direction to go in. The other big factor is the implementation of the dynamics and routines needed to keep the organization moving in that direction and that is achieved with a Continuous Improvement system. This is the way to success: set a good direction (Lean Philosophy, Shingo Model and Toyota Way) and start moving in that direction in a systematic and sustainable way (Continuous Improvement System). In fact, we can only know if there is improvement or not if we define the direction we want to take. Saying something obvious, we only know if we are moving forward if we know which way we want to go.

Many famous stories and phrases refer to this idea. For example, in the story of Alice in Wonderland, at one point Alice asks the cat: "Can you tell me which way I should go?" That depends a lot on where you want to go," replied the cat. "But I don't know where to go!" - said Alice. "If you don't know where to go, any road will do".

Another example is a sentence by the great thinker Seneca: "When one sails without a destination, no wind is favorable".

It is essential to define and believe in a path and to share that same path or direction with the whole organization. Toyota passes this direction to be taken to divisions, departments and sections through the use of the *Hoshin Kanri* practice we presented at the beginning of this chapter.

5.2 Identifying Problems *versus* Solving Problems

Wherever it is, there are always problems because in reality nothing is, or will be, perfect. Sooner or later machines will break down, defects will happen, people will fail and components will jam. In short, there will always be a problem that has to be solved or something that can be improved. It is important that this awareness that there is always a way to do better is shared by everyone in organizations. Whenever someone, in any section or department, does not identify any problem, then that person is not "seeing" well. According to some sources, among which we highlight Brophy (2013), a very interesting maxim attributed to Taiichi Ohno on this subject is:

"Having no problems is the biggest problem"

Thus, the biggest problem occurs when employees in a certain process, section or department do not identify problems in the very areas where they work. When asked, they often reply that they have no problems, that everything is fine. This is both common and intriguing. Why people can easily identify problems in other services and other areas of the organization, but often have difficulty identifying problems in the services/processes they work on every day? One possible reason is that we too easily stop paying attention to what we see every day. In fact, when we start working in a new area, we notice problems, but as time goes by, we gradually stop seeing them. This can happen because in some cases we accept the justifications of more experienced colleagues (who convince us) and in other cases because we simply assume that there must be some reason why things are the way they are. Interestingly enough, many people even though they are unaware of the reason for a certain practice, choose to look for an improvised explanation at the last minute when someone from outside the organization questions them. Ideally, any question presented by an external person should be taken advantage of to be analyzed internally. This practice should be more cherished in organizations.

At one time, in a certain organization, which we prefer not to name, we identified the need to change the product entry position in a machine with automatic feeding. If the product entered in a different position, this would be physically less demanding for the employee and movements without added value would be drastically reduced. In addition to significant ergonomic improvements, this change would result in reduced waste with movements and consequently improved productivity. The person leading the section in question, when presented with the proposed change, argued that the product had to enter this position due to an imposition of the client (who had defined it as the position on the packaging). This argumentation justified perfectly the reason for this practice. We immediately had no choice but to accept the argument, but as soon as possible, we tried to understand the situation in more detail. After talking to the sales department, we found out that the argument was not valid after all, since the position of the products on the packaging was completely irrelevant for the customer. This example shows that some people, even without any kind of "bad faith", can easily make this kind of mistake when trying to justify existing practices, simply because in their subconscious they are convinced that the existing practice is "obviously" the right one.

Another curious aspect is that many organizations prematurely end improvement efforts because they are satisfied with the results achieved. That should not be the way to be. You cannot implement an improvement and rest; you must immediately start looking for another improvement opportunity. Improvement tasks should not be considered as an additional/special part of the job. Improvement should be part of everyone's job, every day and forever. Improvement work should not only happen at pre-defined times or when there are no other tasks to do, for example, when the demand is low or at the beginning of each month. Continuous Improvement is, as the name suggests, a never-ending process that aims to get better and better. Quite often, we hear section managers say that there is not much left to improve in the sections where they work. We also heard organization owners saying that their production lines were optimized and that there was therefore not much left to improve. This is the kind of thinking we need to change, in ourselves and in those around us, for the good of the whole community.

Much is said about problem solving techniques and structured problem solving techniques, such as DMAIC (Define, Measure, Analyze, Improve and Control), A3 reports, Ishikawa diagram, the 5 Whys, or Kobetsu; however, the biggest challenge is not how difficult the problem solving may, or may not, be. The biggest problem is to accept that a problem is in fact a problem. It matters little that a manager goes to training on problem solving techniques if afterwards that same person cannot identify the problems around him. Let us give a small example to clarify this perspective.

> The department where we work is in a building with many doors to the outside. Very often, even in winter, we find these doors open. Since there are heaters running in the corridors all day long, the energy wasted is enormous. As if this were not enough, the areas next to these doors are cold and uncomfortable. To solve this problem you do not need to resort to great problem-solving techniques. In fact, in our department there are many university professors with the technical capacity to find a solution to the problem. Solving this particular problem is easy; what is difficult is for people to assume that it is a problem.

Like this example, there are a myriad of other problems waiting to be identified in organizations, which, once identified as problems, are easy to solve. Look around your organization and see how many problems exist without anyone taking notice. Continuous Improvement is intended to be a systematic and relentless way of solving problems and making the organization better and better. Of course, more complex problems should be tackled in a systematic and structured way, using problem-solving techniques such as those mentioned above.

5.2.1 Systems of suggestions for continuous improvement

It is relatively easy to accept that employees in any organization are capable of identifying problems and opportunities for improvement around them. In fact, it is quite common for employees to complain to co-workers, or even other people close to them, about things that do not work well and even describe how they should work. Then with time, they stop doing it for a variety of reasons, but it is certain that at least for some time they do it. The idea of using this critical spirit and natural creative potential of people to suggest possible improvements is not new. You will all know examples of organizations where suggestion boxes have been placed on walls. Suggestion boxes can be seen around in various contexts, and although they are no longer as dangerous as they were in the time of the eighth Shogun named Yoshimuni Tokugawa (mentioned in chapter 2), they are almost useless in most cases. There are acrylic suggestion boxes, stuck on the walls of many organizations, simply gathering dust. Reality tells us that a suggestion system to work needs much more than just acrylic boxes and formatted forms available to be filled in

Suggestion systems are, however, popular in a very high percentage of organizations with well-functioning Continuous Improvement systems.

Some of them even have award schemes, which can include all suggestions or just the best ones. There are cases where the author of the suggestion receives an economic reward proportional to the gain the organization had with that suggestion. In other cases, however, all suggestions, regardless of the resulting gains, are rewarded with a fixed prize (cash, gifts, organization products, time off, etc.). In addition, many organizations formalize on visual boards and in internal publications, the recognition, before the whole organization, to this or that employee for the quality or impact of his/her suggestions.

Those who have already experienced the creation of suggestion systems have probably learned that the first difficulty is to get the employees to start making suggestions. The second difficulty - far more complex - is to follow up on the suggestions that start coming into the system. This is another issue, which requires careful thought before implementation. To respond to the low adhesion of employees to give suggestions, there is a frequent temptation to give incentives for the suggestions presented.

Another temptation, which is not so common, but exists in some organizations, is to force a minimum number of suggestions (for example, per week) for each department or section of the organization.

A long-standing friend and important manager of an organization of considerable size in the Braga area (I hope you don't mind if you read this book), faced with the low adhesion of his employees in providing suggestions for improvement in their respective boxes, decided to force a certain minimum number of suggestions to be submitted each week. As the organization's culture was not conducive to spontaneity in the submission of suggestions, the employees found themselves at the end of Fridays forced to make up the minimum number of suggestions they were obliged to make in order to go home. As you can imagine, the quality and practical use of most of these suggestions was not spectacular, so this system resulted in a questionable balance between the managers' work in evaluating the suggestions and the poor quality of the suggestions. The idea of this important manager was very interesting because as he was not getting spontaneous suggestions from his employees, he decided to act and make a decision so that suggestions would start to appear. It is important to highlight that the simple fact of wanting suggestions so strongly is already a very revealing sign of the openness of that manager to the involvement of all in the improvement of the organization. Unfortunately, it did not work out as he had planned because the conditions were not created for the suggestions to appear spontaneously.

Regarding the first difficulty, which is getting employees to participate with suggestions for improvement, you have to bear in mind that people do not move quickly from a culture where nothing can be questioned to one where everything must be questioned. There is naturally some apprehension at the beginning about pioneering suggestions. You have to take it easy and accept possible apathy from employees. In addition, it is important to be consistent in the messages that are passed on to employees. Messages such as "respect for the individual", "lead with humility", "do not blame or judge" and others in this vein, should be verbalized and practiced by managers every day. One approach that is often used is, instead of asking employees for suggestions, to ask them to report problems, especially problems that affect them directly or indirectly. Problems such as stress, exaggerated effort being asked of them, lack of clarity in what is being asked of them, inconsistency in the orders they receive, or something unpleasant at work. Employees may be more easily interested in solving their own problems and dissatisfaction with their work than solving the problems of the section or the organization. This is a way to unlock resistance to identifying problems or proposing suggestions. If employees feel that these first demonstrations of displeasure with something are addressed in the right way, in an attempt to solve it, then they are more prepared to come up with further problem identification or suggestion identification.

The second difficulty is related to the ability to respond to suggestions made by employees. It is vital to be able to respond to the first suggestions that appear. This response has necessarily two phases, the phase of decision making on whether the suggestion is implemented or not and the phase of recognition and positive reinforcement. Regarding the decision on whether or not to implement the suggestion it is important to keep in mind that suggestions should be welcomed and whenever possible they should be implemented even if their implementation does not have direct results on the short-term performance. At this beginning of the process, it is most important to create an atmosphere of trust from the employees towards the continuous improvement movement. People should feel that their opinions are heard and that there is a genuine will to involve everyone in what are common goals and where everyone stands to gain.

Suggestions that improve the job satisfaction of employees should be welcomed and encouraged even if at first sight they do not bring any gains for the organization. This is clearly portrayed in the Toyota Way principle "Base your management decisions on a long-term philosophy, even at the expense of short-term financial goals". Another way of looking at these suggestions, which do not generate immediate gains, but which may bring gains in the medium and long term, is to assume that they generate so-called "intangible gains".

> This designation was often used by a famous Japanese consultant who worked for a large multinational organization that has a factory in Braga in the electronics sector.

It is important to note that judging a suggestion for improvement as resulting in "intangible gains" requires experience and a lot of sensitivity in dealing with employees and managers. Anyway, even if there are doubts one should always look at the impact on the employee. One should always think of the possible benefit that the suggestion might bring to the employee's morale.

Insisting on the meaning of intangible gains, we believe that these may be of at least three types. The first is that type of gain that slightly reduces the effort of the employee or slightly reduces the risk to which the employee is exposed (safety or ergonomics) but which does not translate into objective improvements in any indicator. For example, slightly raising the pallet position so that the employee has easier access to the materials. Although there is no doubt that the movement of the operator is facilitated, the reality is that it may not really influence productivity, lead time or any other indicator. The second type occurs when the proposed change has no immediate effect but will lead to an improvement in the future. Let us look at an example. Imagine that in a sequence of jobs it is possible to slightly reduce the time of one of the intermediate jobs that is not a bottleneck. This small reduction in the time of the intermediate post changes nothing overall and therefore has no objective benefit. Sometimes, however, this change provides opportunities for future changes that will result in improved performance. Finally, the last type of intangible gain presented here is what happens when the change does not bring any improvement but the employee is satisfied and motivated to be taken into account. Just the fact that we make an employee satisfied is already worth it, but in fact it goes a little further than that because this employee is motivated to make other suggestions that may bring real gains.

There is a strong idea that should be kept alive in organizations with Continuous Improvement systems. Everyone should continually look for changes that might make the job easier or more enjoyable if only a little bit. Employees should be encouraged to suggest improvements that are often overlooked, such as:

- Putting tools and materials a little more "at hand",

- Move equipment a little closer together to reduce travel,
- Simplify the forms of communication with colors, lights, and other types of signs,
- Clarify and standardize operations,
- Improve lighting and cleanliness in work areas,
- Increase task diversity to avoid overload and repetitive operations,
- Improve as much as possible the comfort in the workspace.

One should encourage the kind of suggestions that bring insignificant gains but if, when there are many and over time, result in huge gains for the organization.

Notice that the approach of many low impact improvements implemented constantly over time compared to the approach of few high impact improvements implemented sporadically is similar to the relationship between continuous (or one-piece flow) production and batch production. The concept and principle of flow also makes perfect sense in continuous improvement and a bit in our lives.

References

Akao, Y. (1990). *Hoshin Kanri: Policy Deployment for Successful TQM.* New York: Productivity Press.

Brophy, A. (2013). *The Financial Times guide to lean : how to streamline your organization, engage employees and create a competitive edge.* Pearson.

Liker, J., & Franz, J. (2011). The Toyota Way to Continuous Improvement: Linking Strategy and Operational Excellence to Achieve Superior Performance. McGraw-Hill Publishing.

Rother, M. (2010). *Toyota KATA*. McGraw-Hill Education - Europe.

Womack, J., & Jones, D. (1996). *Lean thinking: Banish Waste and Create Wealth in Your Corporation.* New York: Fee Press.

6

Maturity of Continuous Improvement

This chapter tries to explore the concept of Continuous Improvement maturity in an organization. It is presented an evaluation proposal with four maturity levels, the lowest being called occasional improvements and the highest being total continuous improvement. This proposal of four maturity levels has the objective to allow any manager interested in Continuous Improvement to identify in which maturity level his organization is, and in the light of this finding, he will be able to decide which kind of transformations and/or implementations will be necessary to go up a level.

Our experience in working with organizations in our region has led us to develop a great respect and admiration for entrepreneurs in general. They have undoubtedly been the drivers of our economy but they are not always appreciated. Some of them have had far fewer opportunities to educate themselves than their employees and yet they are there enabling others to enjoy the opportunities that many of them did not have. They have a huge challenge ahead of them, the challenge of seeking to evolve into determining the right vision for the business and the operation, and the challenge of building the right culture of rigor and discipline, with humility and with respect for people. The success of continuous improvement systems depends essentially on that person leading the organization. He or she must lead the movement otherwise it will not work. Any Continuous Improvement (CI) consultant quickly learns that if the main leader of the organization does not understand the idea and does not take it on board, it is not even worth the effort. Middle managers can also be blockers in the development of Lean CI approaches but it will be easier to overcome this difficulty.

Small and medium-sized organizations usually use external consultants to implement CI systems, with a great deal of involvement from these consultants in the early stages but becoming less frequent as the maturity of the CI system increases. Some organizations, however, build their CI systems with internal resources only. In these cases, there has to be at least one person within the organization with leadership power and great understanding and focus. In large multinational organizations, with factories spread all over the world, the scenario is not very different. Here too we can find cases that use external consultants for the development of their own CI systems and cases where these systems are developed internally.

In this book, it is proposed that the direction organizations should take is towards organizational excellence, as understood by "Lean thinking or philosophy", the "Shingo model" or the "Toyota Way". It is true that these models are all different, but they all share the same source which is the approach developed at Toyota that has been extensively referred in this book. These models are called Toyota Inspired Excellence Models (TIEM). Nevertheless, many practitioners and academics in the field assume that the term "Lean" carries with it everything developed at Toyota and not only what was defined by the creators of the term and authors of the famous books "*The Machine that Changed the World*" (Womack, Jones, & Roos, 1990) and "*Lean thinking: Banish Waste and Create Wealth in Your Corporation*" (Womack & Jones, 1996). Therefore, when we simply use the term "Lean" we are including the whole body of knowledge that includes the Shingo model, the Toyota Way and Kaizen. That said, a transformation or change in an organization is only considered an improvement if it moves the organization towards its objectives and strategy. If the organization in question wants to follow the principles of excellence according to "Lean" (or TIEM) then it is clear which direction the CI system should focus on. This "Lean" vision can be described in many ways, including for example: customer focus, customer-driven production, the goal of one-piece-flow, the constant reduction of the 7 types of waste and the use of the full human potential, leading with humility, having a long-term philosophy, etc. (Figure 6.1).

Much of what is at the heart of Lean concepts and principles is actually counterintuitive. One example is the principle of seeking to improve product flow and pull production. This principle involves reducing batch sizes in order to produce only what is required by the next process and not more than is needed and not earlier than the time when it is needed. Production in large batches is seen very intuitively by most people as being more effective, more efficient and cheaper, than producing small quantities at a time or even one unit at a time. One possible explanation, suggested by James Womack for this intuition, is our very agriculturally connected past over the last tens of thousands of years. In agriculture everything is done in batches. Sowing is

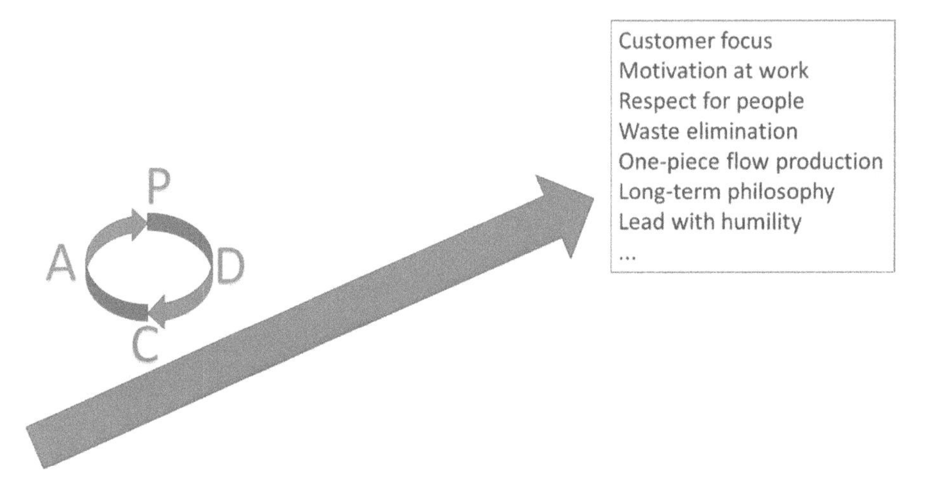

Figure 6.1 Continuous improvement with a vision of excellence in organizations.

done all at once with all the seeds, the weeding process is also done over the whole planted area at once, the pest removal process too, then the harvesting is done all at once, the eventual drying of the grain too and finally everything is stored. This legacy where every process is carried out on every unit of the product at every stage may have left imprinted in our "subconscious" this tendency for batch creation.

> To assess the maturity of CI in organizations, normally organizations start by applying Lean concepts and principles in the physical production areas and only then eventually start including the administrative and support areas. We only know one example of an organization, a public waste treatment organization called Lipor, located in the Porto area, which, unlike all the others, started its Lean journey in the administrative areas. Only a few years later, due to the good results achieved with the experience in the administrative areas, they decided to move to the production area.

It is not easy to define a taxonomy of maturity levels, since the boundary between levels can be somewhat fuzzy; but it is always possible to try to establish some patterns. After some attempts (involving more or fewer levels), it was possible to identify four major levels for the maturity of KM in organizations that will be described below and that are the following: Level I - Occasional Improvements; Level II - Routine Improvements;

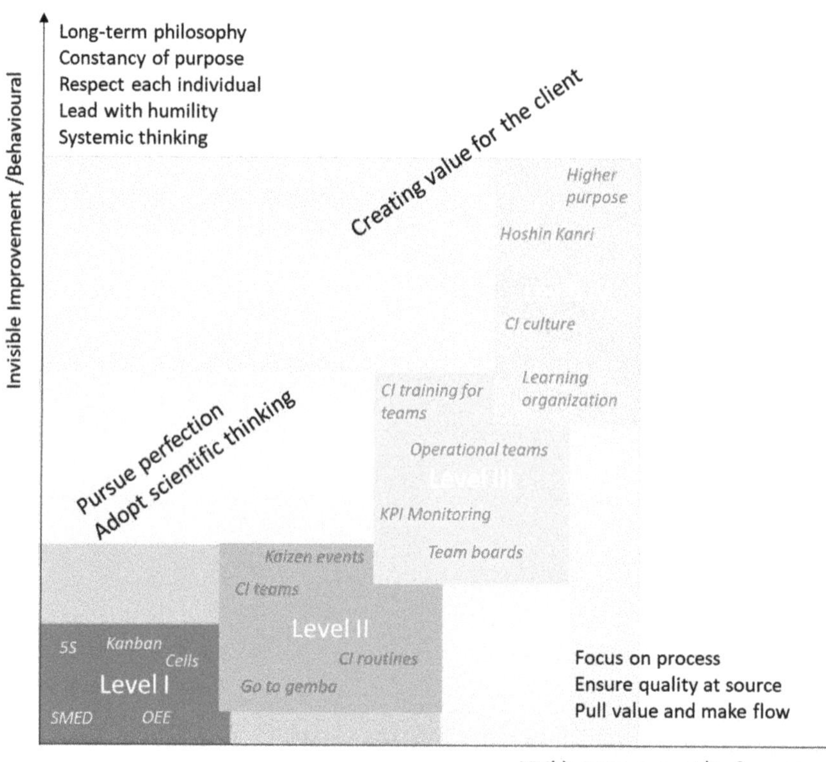

Figure 6.2 CI Maturity levels.

Level III - Structured Continuous Improvement; and Level IV - Total Continuous Improvement. The representation of these four maturity levels regarding their impact on the visible and invisible performance (behaviour/culture) is presented in Figure 6.2.

6.1 Level I – Occasional Improvements

At the lowest level of the scale (see Figure 6.2), organizations only carry out improvements in an occasional way involving external consultants or not. The reason why this level was assumed here is that on the one hand many practitioners and academics call this kind of occasional improvement "continuous improvement" and on the other say that it is the first step for an organization to enter into a process leading to CI itself.

To call a CI maturity level "occasional improvements" is even inaccurate because, in fact, at this level there is no continuous improvement.

It is important to clarify that in many cases the designation "continuous improvement" is not even used because there is some confusion between the concept of "continuous improvement" and the concept of "Lean", as it was already mentioned before in this book. It is even much more frequent that organizations want to "implement lean" or "implement kaizen" than to implement "continuous improvement". There can be many reasons that can lead an organization to start this process and here are some very frequent examples:

- One of the organization's managers had been to a lecture on Lean or visited an organization that had interesting implementations in that area and thought it was a good idea to try it out;
- An important client started to suggest or "force" the introduction of Lean tools in the organization;
- A new manager was hired and has experience in the subject and started to convince colleagues to try it,
- An Engineering or Management trainee started to try to introduce some lean tools in the organization.

Any of these examples, like others, can be the "seed" or the first step to start including the Lean philosophy and the LC in some routines of the organization. It is not too much to remember that a significant part of the cases that start this way do not result in success but some end up resulting in success with huge gains for the organization.

Whatever the reason to take the first steps on the Lean path, organizations typically start with projects such as: implementation of 5S (*a workplace organization technique*), global value stream analysis using VSM (*Value Stream Mapping*) and including the implementation of some changes in the flows, SMED (*Single Minute Exchange of Die*) study on a machine with a long setup time, OEE (*Overall Equipment Effectiveness*) monitoring on some important machines, changing the position of equipment to reduce transport and handling waste, and many other options.

The implementation of 5S is, for a variety of reasons, one of the first tools to be adopted by organizations when they start on the Lean path. Implementing 5S is a very popular way to start the Lean journey for various reasons but probably because it results in visible changes in the work areas while generating benefits in the organization of the workstations and the work itself. Spaces become more organized, materials are easier to find, safety and quality of work increases. The start of a 5S project motivates employees and managers, in most cases, but the real challenge is to keep 5S alive inside the organizations, which is actually the essence of the fifth S. It is vital to assume that the implementation of 5S is like a long distance race and therefore a lot of discipline is needed. It requires a paradigm shift and in the culture of an organization for 5S to endure over time. This cultural and behavioral change either happens or else all the initial effort in implementing 5S will be lost. Let us look at an example to clarify the idea. Marking a hallway with yellow tape taped to the floor is a task that requires some dedication but is far less challenging than keeping the aisles, marked with yellow tape, free of materials, carts and pallets, parked in them every day. If leaders do not understand the power of compliance then this type of project will never succeed. If the rules are not adequate then they need to be redesigned to make them more appropriate, but once they are defined, leaders have the responsibility to enforce them until they are changed.

The cultural evolution to comply with rules is very important, but no less important is the cultural evolution to make rules that you want to enforce. There are rules that are created in our organizations and posted on the walls that seem to have already been put there knowing that they would not be complied. Sometimes by suggestion of consultants others by legal impositions such as security issues. How many of our readers have entered a section of a factory where it is compulsory to use ear protectors, the sign is there, but few comply? The same goes for the mandatory use of goggles, gloves, etc. One of the problems is that not everything that is said to be complied with is actually complied with. There is a silent understanding of the rules behind the rules. We all need to grow into this theme of compliance for the sake of our future as an economy and as a society.

An organization located here in the north of Portugal, belonging to a multinational group, which in a very difficult period of survival, around 2004, had to replace the production director because he could not align himself with the principles of lean production. The top management had taken the decision to pursue this lean path and only after replacing the production director did they manage to make the necessary changes so that the course would be aligned with these principles. In this way, they managed to survive a very difficult period and ensured that the organization would not be relocated to another part of the world, as this would be its fate if it could not become competitive. This organization not only survived but also absorbed other companies that were located in other parts of the world, creating jobs directly and indirectly as a multiplier in the supply chain.

Some organizations, because they met a context conducive to success with these first implementations, evolved to the next level of maturity. On the other hand, there is a huge number of other organizations (*it is not easy to say what percentage simply because there are no studies on the subject, only some speculations*), where these initiatives failed for a variety of reasons and so they gave up pursuing this Lean path. The main reason is that the number one person in the organization, the main leader, did not believe in this path or simply could not understand it.

6.2 Level II – Routine Improvements

At level II, organizations are not limited to occasional improvement projects but have routines established and resources allocated to introduce improvements. For this to happen there will have to be at least one person (*it is desirable to be a small team*) with the responsibility of promoting the continuous improvement in the organization. When an organization assigns someone the responsibility of promoting CI it is because there is already some formal position from the top management towards that path. When organizations are at this level it is more difficult to give up the CI movement but nothing is guaranteed. It is important to mention that giving the responsibility for implementing Lean or CI to a single person is often frustrating for him/her and as a consequence for the organization itself because results hardly show up. That person will probably find resistance from many middle managers, foremen, and workers who will use arguments such as: "we don't have time for that", "that works well in organizations with money but it doesn't work here", "we tried it before but it doesn't work", "that's very nice but we already have a lot to do", and other arguments like that. Then the business owner charges that person for results that do not appear and, unfortunately, "Lean" fails.

In cases where only one person is given the responsibility of achieving CI, what actually happens in the best-case scenario is that that person manages to make some disruptive improvements. That CI animator somehow identifies, with or without systematic criteria, some improvement opportunities and applies some "Lean" tools. As in the previous level, examples of these disruptive improvements can be: A layout change with better flows; a reduction of the setup time of a machine by applying the SMED (*Single Minute Exchange of Die)* technique; some changes in procedures; applying 5S in an area of the factory; etc. However, some of these disruptive improvements run the risk of gradually regressing to the previous state. Such cases are frequent, for example in the implementation of 5S, where over time everything returns to the previous state where nobody worries about putting the tools in their places or keeping the spaces clean and organized. This type of phenomenon can result in the discredit of the Lean approach and consequent abandon when the organization leaders are not convinced that this should be the way forward.

A similar case occurs when the organization hires an external consultant to "implement lean". Quite often the leader does not have the time to spend with the consultant but hopes that he or she can achieve improvements without disrupting people's routines too much. The results are similar to those achieved in the previous case. As the main leader does not have time to spend on these transformations, sometimes the consultant is assigned an internal privileged contact who may be a person without competencies in those areas of Lean and CI. The result is often failure. In other cases, when leaders have relatively well understood the long-term benefits of Lean, then they try again in another format with more human resources until some transformations take place on a sustained basis.

Assigning CI responsibilities to a team is a much more promising thing to do than assigning them to one person alone.

An example we know clearly at maturity level II with CI being assigned to a team is that of a car assembly company we visited in the year 2003. That was a long time ago. That organization had a team of engineers who were dedicated only to CI tasks. That team visited cyclically all the sectors of the line in order to analyze in detail which problems and/or improvement opportunities they identified, and which wastes they managed to reduce. Once the improvement opportunities were identified, the team studied solutions and implemented those solutions that brought improvements in productivity, quality, ergonomics, safety, etc. Then they moved on to the next sector of the line, and so on until they had analyzed the whole line. When they had finished analyzing all the sectors of the line they would go back to the beginning again. There was actually a system in place to ensure the sustainability of continuous improvement. This is clearly an example of CI maturity level II.

Another type of approach that clearly fits this level of maturity and still exists especially in many multinational organizations is one that focuses on so-called Kaizen events. These events are carried out periodically and are often led by external consultants or experts from the parent organization who travel cyclically to each of the group's factories. These are disruptive transformations in large areas of production or in specific processes. Typically, during 3 to 5 days, a team led by a specialist analyses in detail a certain sector of the factory and implements actions aimed at achieving the performance objectives assigned for the event. There is a standard methodology to be followed in this type of event and several books have been published on the subject (Martin & Osterling, 2007)(M. Hamel, 2010).

Another approach, slightly better than the previous one, is the one where a team is formed to dedicate itself to the management of the continuous improvement. This is a different case from the previous example about the team of engineers doing field implementations. Now the case is the creation of a team of management, animation, promotion of the CI but that does not necessarily has to have the responsibility to carry out implementations on the field. When such a team is created in an organization, it is already clear that the organization gives more importance to the CI. It is normally expected that such a team will initially look into the Gemba[1] to identify opportunities for improvement. These improvement opportunities are transformed into projects or events *(the so called "Kaizen events")* that lead to disruptive improvements such as changes in the equipment layout, implementation of 5S, introduction of new technologies, creation of standards, reduction of setup times of important equipment, etc. This approach to CI does bring benefits to the organization but usually either it has evolved into a more serious approach to CI or its impact may be lost. As the improvements implemented result in smaller and smaller gains, this team can be called into question. It is desirable that this team gains skills and experience in continuous improvement and Lean, as well as in other models for excellence in organizations, so that they evolve to gradually develop and implement practices that involve more and more employees and managers in the continuous improvement effort.

6.3 Level III – Structured Continuous Improvement

Level III, is called Structured Continuous Improvement, being the maturity level in which the CI movement has already transformed the organizational structure of the organization. At this level, a large part of the organization is already organized by teams of collaborators as units with responsibility that besides the operations on products or services, also have CI routines. These

[1] Gemba is a Japanese word. Saying "go to the Gemba" is equivalent to saying "go to the field", i.e. go to the place where the operation takes place. The idea is to go to the field, see for yourself, and talk to the people on the ground rather than simply looking at reports about what is happening on the ground. This concept although simple is incredibly effective.

teams of employees may be called "natural teams" or "operational teams". The term "natural teams" is used by the Kaizen Institute with the assumption that these employees already naturally form a team because they work in the same physical area, on the same type of process or the same product family. In essence, they are already linked in their work routines. These operational teams are assigned short and medium-term objectives and goals to be pursued. These objectives are often assigned by consensus among the team members so that everyone feels responsible and motivated to achieve them. In many organizations, there are forms of recognition when teams achieve the objectives or because they have made a significant improvement in their area of work. Some people publish in the common areas of the organization a photo with the identification of the team of the month, for example, as a way to recognize some merit in a team. The recognition of individual people is not very advisable when teamwork is intended, but this subject is far from obtaining a generalized consensus.

Natural or operational teams are trained in simple continuous improvement routines, which include for example: skills to maintain 5S routines, skills to monitor their KPI's, skills to identify small problems and improvement opportunities in their work area and in some cases problem solving skills. In some organizations, these teams of employees are only expected to follow the standards and achieve stability in their daily work. In other organizations, these teams are also expected to identify problems and improvement opportunities at their workplaces but are not expected to implement them.

Another entity that is typically present in CI systems is the team that fosters and supports CI. That team is usually composed by a small number of elements with management responsibilities and at least some elements with great experience in continuous improvement. They are usually called support team, or CI management team, and they are responsible for the creation of the natural teams, organizing them and making them grow. In addition, this support team defines objectives and makes decisions regarding possible disruptive projects outside the scope of the natural teams.

Project teams are another type of team that are formed by elements that can be from different departments and that are created to solve a specific problem or produce disruptive improvements within the organization. These teams exist for the duration of the project and are obviously dismantled as soon as the project ends. For their work to be effective it is important that some of their members have skills in structured problem solving, namely the ability to use tools such as A3 reports, Ishikawa diagrams and Kobetsu. It is also desirable that they have some skills in agile team and project management such as Scrum and kanban. This type of team may frame their projects as "Kaizen events" or they may be longer term and less disruptive projects. It can be a SMED project or the restructuring of factory flows, changes in the layout of equipment, restructuring of internal logistics, or any other project that produce significant changes to the way of working. It is usually up to these project teams to improve and update the working standards that result from their projects.

In organizations at this level of maturity, there is however, a certain separation between the normal management and operation tasks and the tasks related to CI. Trying to clarify what is meant by this is that most employees and managers, in organizations at this level of maturity, carry out their daily tasks a little unaware of what is going on in CI. Some people are concerned with operation, management, and others with continuous improvement. There can even be occasional conflicts of interest between area leaders and those responsible for CI. The production director, for example, may not recognize the work done by the CI team. There are even cases where some managers boycott CI work. This problem ultimately will have to be solved by the main leader. When that main leader, the business owner, for some reason is unable to do anything (*sometimes it is family members*) then the CI work is seriously compromised.

In cases where there is an increase in demand or, for some other reason, the production rate is asked to increase, at this level of maturity the continuous improvement effort ceases to exist. It is a bit like that maxim "in wartime you don't clean weapons". All the routines and norms of continuous improvement, or at least a large part of them, are no longer carried out because what is important is to produce. Then you begin to notice the relaxation in the operations associated with continuous improvement. In relation to 5S you start to leave tools out of place, pallets with materials are forgotten in the corridors, pallet racks are left just anywhere, etc. Updating the teams' KPIs is also left for later, then for the next day and gradually stops. Even detailed work instructions are no longer followed with the connivance of the leaders. There is no more time for CI meetings, for work planning and for generating ideas for improvement or problem solving. In many organizations, even multinational organizations operating in highly competitive markets, with large numbers of skilled and competent industrial managers and engineers, we see this happening. Many of them are even recognized as benchmark organizations in some sectors, but suffer from this type of behavior.

These problems happen because these organizations have not yet fully integrated CI in their DNA. Although visibly it seems that there is great maturity in CI, in fact the invisible part is not working. In these cases, top management has not yet taken on CI as part of the nature of the organization.

Although the message is politically correct about CI, day-to-day actions, especially when pressure builds, do not reflect the formal, written messages.

6.4 Level IV – Total Continuous Improvement

An organization will be at this level of maturity, which is here called "Total Continuous Improvement", when continuous improvement is part of the normal operation of everyone in the organization. Furthermore, for an organization to be recognized at this level of maturity it must also have a vision aligned with other important principles inherited from Toyota such as respect for the individual, leadership with humility and the existence of a higher purpose. In Figure 6.2 we can see the relative position of each maturity level in relation to two axes, which represent the two main worlds of excellence in organizations. On one hand we have the axis that represents how much the organization is evolved in the visible part of excellence in organizations, and that can be evaluated through performance indicators (KPI), and on the other hand how mature the organization is in the invisible part (behavior, attitude, culture).

Mike Rother refers to this division between the visible part, related to tools and techniques, and the invisible part, related to routines, attitudes, behavior and culture in his book Toyota Kata (Rother, 2010). The visible part has been more easily replicated because it is easier to notice, although often without success because the invisible part is neglected. The invisible part is difficult to copy and is often neglected by many managers because they believe that getting the right technology and applying the right tools is enough. This kind of division is also referred to in other ways as for example in the interpretation of operational excellence developed at the Shingo Institute making the division between KPI (Key Performance Indicators) and KBI (Key Behavioral Indicators) (Edgeman & Barker, 2019).

It can be said that this invisible side of organizations or the behavioral and cultural side is as present and vital in the success of organizations as it is neglected by most of their managers and leaders. On this subject, Professor Edgar Schein of MIT School of Management makes the following very interesting observation (Ahmed & Saima, 2014):

> *"The only thing of real importance that leaders do is to create and manage culture. If you do not manage culture, it manages you, and you may not even be aware of the extent to which this is happening".*

In the graphical representation of Figure 6.2, we can then see the relative position of the four maturity levels in the two axes considered (the visible side and the invisible side). We consider that at the lowest maturity levels, more emphasis is given to the visible side of the CI and as the maturity level rises, the greater is the manifestation of the invisible side (behavior and culture). It is at this higher maturity level that the existence of a higher purpose, the focus on the development of people and the CI, being embedded in the day-to-day management and operation of the organization, is best manifested. CI tasks or routines are an integral part of one's job.

At this level of maturity the quality of the leadership of the organization plays a key role, as the vision, purpose and strategy will need to include continuous improvement very clearly and get the message through the organization very well in order to create the right culture. The leaders of the organization define and effectively communicate to the whole organization their long-term philosophy, as well as the defined vision and purpose. As noted earlier, many successful organizations define "a higher purpose" than just profit or other "self-centered" purposes. Altruistic purposes, with a focus on the common good, the future of the planet, etc., convey ideas that are usually well received by employees, shareholders, customers, suppliers and the community at large. It is very different for an employee to see or identify themselves as putting blocks on walls or to see or identify themselves as contributing to a building that will change the world or significantly improve people's well-being. It is also very important at this level of maturity to convey effectively to the whole organization that long-term goals are more important than short-term goals. There may be short-term performance losses for the benefit of long-term goals and objectives.

The objectives and goals defined to be achieved in certain periods are shared to each unit of the structure, which creates its own objectives and goals in line with the organization and according to its own context. In turn, each unit communicates to all its sub-units these objectives and goals and so on according to the *Hoshin Kanri* principles presented in chapter 6. In this way, the whole organization will be aligned with the top management or leadership of the organization.

An analogy can be established between what happened two decades ago with the quality area and what is happening today with CI. The quality area in the past was the responsibility of the quality director and production was the responsibility of the production director. The quality controllers took orders from the quality director and the operator took orders from the production director. The objective of production was to produce as much as possible and the objective of quality was to separate, as best as possible, the good products from the defective products. Quality was a layer of operations and bureaucracy above production but without interfering practically in production. That approach unfortunately still exists in many organizations and in many organizations. In today's approach, adopted by many successful organizations, operations and quality are closely linked. The truth is that they are linked by nature. The defect is created in the operation and so is the perfect part. The idea of getting it right first time and assuming that the operator is responsible for the quality of what they produce is the right approach. Instead of asking the operator to produce and asking a controller to control we just ask the operator to produce while doing the control and ensuring that the part produced is good. Of course, we cannot always avoid a control to be done after the operation or operations but the principle to follow whenever possible is to guarantee that it is done right the first time.

In the same line of reasoning, it can be expected that it will be a long time before, in most organizations, CI routines and obligations become a natural part of everyone's daily work. There are some organizations, Toyota being a good example, where CI is increasingly more integrated in everybody's tasks and obligations, not being a responsibility assigned only to some. CI is more and more a part of management and operation. In this sense in excellent organizations or organizations with high levels of excellence, it is expected that operation, quality and CI are present in all processes and in all people.

Regarding the role of people, in highly successful organizations, it is common to have a great emphasis and investment in the invisible side, in the concern with people and in the recognition of their infinite potential. It is important to keep in mind that people are the only resource in organizations with infinite potential. This potential does not refer to the physical capacity to execute operations but to the infinite capacity to create and develop solutions and alternatives for the organization to become better. When an organization has its CI at the highest level of maturity it is because that organization understands very well the infinite potential of its people. It is in this resource of infinite capacity that organizations should invest the most, as it is in this resource that the infinite earning potential lies. When an organization creates conditions for its human resources to grow and develop as people, it is exponentially increasing its capacity to grow and succeed as an organization. This is because the greater the value of people in their capacity to create solutions and alternatives for improvement within the organization, the greater the value of that organization will be. Along this line, we can add a positive feedback effect because the greater the development of people, the greater the development potential of the organization, which in turn results in greater investment in people, and so on.

The result will be a curve somewhat similar to the curve presented in Figure 6.3. As the organization moves towards excellence and continuous

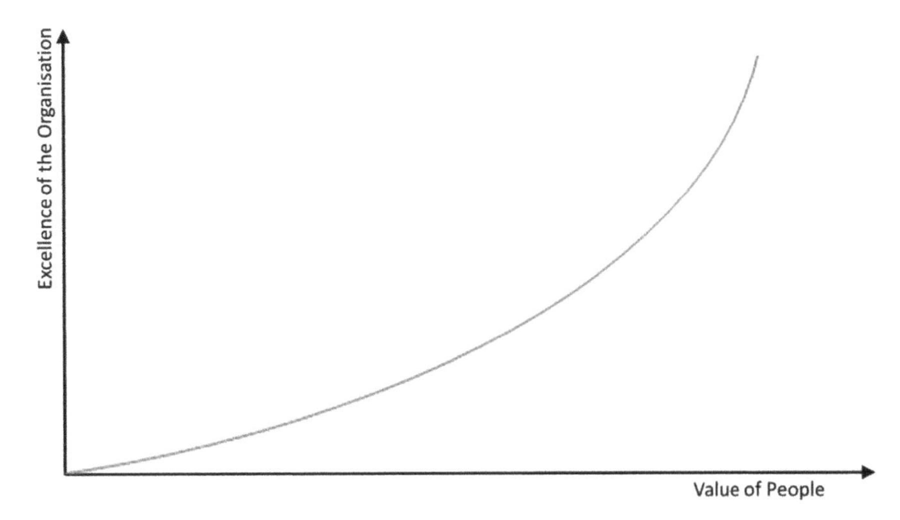

Figure 6.3 The value of people and the level of excellence in organizations.

improvement the more it will grow its people, and at the same time the more people develop, the greater the improvements and the higher the level of excellence. There must be a moment when this process becomes automatic and natural (positive feedback) and it will not be easy to say exactly when it starts happening. In our opinion, this happens when the organization reaches maturity level IV or total continuous improvement. We have to bear in mind that this state does not last forever, and that the cyclical positive feedback effect may suffer disturbances that can break it. The dynamics of economies, societies and technology itself can wreak havoc even on organizations that are in this state of cyclical positive feedback; indeed, nothing can be taken for granted.

6.5 Resources Spent on CI

In terms of the materialization of CI, at the highest maturity level, improvement is part of the normal day-to-day work at all levels of the organization. At this level of maturity, everyone in the organization routinely devotes a small part of their daily time to CI tasks. There is no fixed rule or scientific knowledge about the ideal percentage of time that on average should be devoted to CI by the human resources of an organization but it would be good if there was some reference about it. Not spending any time on improvement tasks is bad but spending all the time on improvement tasks is also bad, as work needs to be done so that products are delivered and services are provided. Many people say that they do not have time for continuous improvement. About this argument, a funny metaphor is often told to clarify the idea:

A woodcutter was chopping down trees uninterruptedly with his axe and you could see he was very focused on that. His aim was to cut down as many trees as he could in the time he had. There is however, one piece of evidence that needs to be attended to, is that the more he cut the less effective his axe was at chopping. Another man watching him approached and suggested that the woodcutter sharpen his axe. The woodcutter replied that he could not sharpen the axe because he had no time for it.

Stephen Covey also refers to this well-known story in his famous book "The 7 habits of highly effective people" (Covey, 2004). In this book, the seventh habit proposed by the author, which is called "tuning the instrument", is directly related to the metaphor of the woodcutter and also aligned with the principle of continuous improvement.

Returning to the question of what should be the optimal time spent on CI operations, or time spent sharpening the axe, there is a famous metaphor attributed to Abraham Lincoln which is as follows (Foussard, 2020):

> *"Give me six hours to chop down a tree and I will spend the first four sharpening the axe"*

In this sentence, it is suggested that more time should be spent on CI tasks than on the actual job tasks of adding value to products and services. This is not actually the practice in organizations. The percentage of time spent on CI is far less than that. Although publications that provide an indicative value are very rare, in the book "*Lean math: Figuring to improve*" written by Hamel & O'Connor (2017) there is an indication of the value of 3% as a reference for the percentage of human resources that should be dedicated to CI tasks in an organization. If we assume that having 3 out of every 100 human resources (HR) dedicated only to CI is more or less equivalent to having all the organization's HR dedicate on average 3% of their time to CI tasks, then we have here an interesting value as a reference. Now, 3% of human resources dedicated to CI should be more expensive than 3% of all human resources' time simply because (by assumption) human resources dedicated exclusively to CI, with CI knowledge and able to foster CI have higher salaries than the average employee. This may be debatable but in most organizations this is the case. The human resources who are dedicated to fostering CI are usually middle or senior management and therefore they often have higher salaries than most employees.

What value in time are we really talking about then in relation to that 3%? Considering 460 minutes as the time each person in the organization spends per day on his/her work, we would have 13.8 minutes as the time each person should spend on average on CI tasks. This value as an average seems to be realistic since some management positions will have to devote more than this amount of time per day while most employees may use a little less. Daily 5-minute meetings of operational teams represent a fairly common practice in organizations with organized continuous improvement systems. If we add up those 5 minutes per day of all the employees with some occasional longer weekly meetings, and if we also add up the time spent in occasional continuous improvement events and the time spent in managing and supporting the continuous improvement then we easily arrive at the 3% figure mentioned.

6.6 Indicators of the CI System

The quality and effectiveness of an organization's CI system can be evaluated in a variety of ways. One way is to measure its impact on overall organization performance. If the organization is reducing costs, improving service, improving productivity, improving quality and improving employee motivation, then we can say that the CI system is effective. It will be as effective as the improvements achieved in the results. Nevertheless, it is advisable (and it works effectively) to create and pursue specific indicators of the CI system itself. With them, it is easier to establish the desired correlation between the evolution of these indicators and the evolution of the organization's global performance indicators.

The number of improvements implemented per person per month or year is an interesting indicator of the vitality of the continuous improvement system. There are even some indicative references so that the organizations can compare themselves. According to a book by Joakim Ahlstrom entitled "*How to succeed with continuous improvement: a primer for becoming the best in the world*", published in 2014 (Ahlstrom, 2014), 13 improvements implemented per person per year is the top value in European organizations while the best organizations in the world manage to approach values close to 20. We suggest the reader, if he works in an organization, to check how many improvements are implemented per person per year in that organization. Most organizations do not do their registration or accounting but even if they did, it is unlikely to get a figure above one improvement per person per year.

There may however be some doubt that the judgement on what should be considered an improvement may be somewhat arbitrary. What should be the criteria for a change to be counted as an implemented improvement? This subject has no direct answer but we can make some considerations about it. In our opinion and according to our experience in the field, a change that is implemented should count as an improvement if the person or team of people that takes the decision considers it as such. The decision maker makes a judgement on whether or not that change results in an increment to the path defined by the organization's vision. If it is understood that this increment goes in the direction of the organization's vision then it should be considered an improvement. It should be noted that all improvements should be formally registered and that this registration should be easily accessible to all for future recall. It is important to highlight that even if the increment is small it should be assumed as an improvement because the most important thing is that everyone makes small improvements as often as possible. Finally, it is important to mention that it is more important to have that spirit, habit and availability of people to make improvements than the judgement of whether or not it should be an insignificant improvement accounted for.

Examples of small real improvements that, although they result in small increments, should be considered as improvements are:

- A number of tools that were mixed up in a box, requiring the operator to bend down under a workbench to look for them, are now in a small, easily accessible drawer on the worktop. This small improvement makes the job easier and reduces the time needed to fetch and put the tools back in place.
- Define and mark the place to park the pallet truck. It is an improvement in line with the 5S tool and reduces the time spent looking for the pallet truck.
- Clearing and marking the area around the extinguisher. This is an improvement in the area of safety.
- A yellow fluorescent cylinder was created that is placed vertically so that it can be seen from afar to indicate the existence of orders for a certain operation. In the past, an employee would have to walk some 80 meters a few times a day to check if there were any orders, which was often not the case. This way the employee no longer has to make the trip in vain and the request is answered more quickly.

The reader will understand that the list could be expanded to take up as many pages as this book but the idea was just to clarify what kind of improvements can be made that are usually overlooked. Each one alone will not make much difference but hundreds of them over a year do. I believe you will understand this will make a big difference to the overall performance of the organization and the well-being of the employees.

References

Ahlstrom, J. (2014). *How to succeed with continuous improvement: a primer for becoming the best in the world*. McGraw-Hill Education.

Ahmed, M., & Saima, S. (2014). The Impact of Organizational Culture on Organizational Performance: A Case Study of Telecom Sector. *Global Journal of Management and Business Research: A Administration and Management*, *14*(3).

Covey, S. R. (2004). *The 7 habits of highly effective people*. New York: Free Press.

Edgeman, R., & Barker, S. (2019). *Complex management systems and the Shingo model: foundations of operational excellence and supporting tools*. Productivity Press.

Foussard, J. (2020). Julien Foussard analyzes the quote by Abraham Lincoln. Retrieved April 14, 2021, from https://julien-foussard.com/en/2020/01/29/julien-froussard-abraham-lincoln/

Hamel, M. (2010). *Kaizen Event Fieldbook: Foundation, Framework, and Standard Work for Effective Events*. Society of Manufacturing Engineers.

Hamel, M. R., & O'Connor, M. (Writer on mathematics). (2017). *Lean math: Figuring to improve*. Dearborn, Michigan: SME.

Martin, K., & Osterling, M. (2007). *The Kaizen Event Planner: Achieving Rapid Improvement in Office, Service, and Technical Environments*. CRC PRess.

Rother, M. (2010). *Toyota KATA*. McGraw-Hill Education - Europe.

Womack, J., & Jones, D. (1996). *Lean thinking: Banish Waste and Create Wealth in Your Corporation*. New York: Fee Press.

Womack, J., Jones, D., & Roos, D. (1990). *The machine that changed the world*. New York: Free Press.

7

Practices to Materialize Continuous Improvement

Although often confused with the Lean, Shingo and Toyota Way models themselves, Continuous Improvement (CI) is actually only one dimension of those models, being referred to explicitly in at least one principle or implicitly in others. This chapter is dedicated only to the practical aspects of implementing CI in the structure and routines of organizations. In this sense, three models for the sustainability of CI in organizations (Toyota Kata, Kaizen and Scrum) are presented, which are available in publications (books, articles and websites) and will serve as reference models in the following chapter. In reality, there is not much literature on how to sustain an effective set of CI entities, artefacts and routines. Many books and articles refer to "continuous improvement" in the title, or in the keywords, but in fact, what they really present are simply tools that are used for eventual improvements in one context or another or in one sector or another. With the more and more frequent implementation of improvement actions, the organizations felt, in a natural way, the need to create mechanisms/structures/systems/routines that allow them to maintain a continuous movement of improvement in all their sectors and involving all their people.

7.1 Toyota Kata

The Toyota Kata model is one of the three continuous improvement models that will be taken as a reference in this book. Toyota Kata (TK) was introduced by Rother (2010) and is intended to represent what the author interpreted as how Toyota maintains continuous improvement in its daily routines.

> The Toyota Kata community, led by Mike Rother, is very open to sharing knowledge, providing a lot of content on the TK model website (Rother, 2020). We acknowledge his sharing attitude for the good of all.

The word "Kata" is used to denote an exercise consisting of several specific movements of a martial art, especially a prescribed pattern for defence against several attackers, used in judo and karate training. The word "Kata" was chosen especially in order to convey the idea of routine, of an automatic pattern of behavior. In this model, it is intended that continuous improvement is so routine that it is part of the behavior on "autopilot", i.e. that it is permanently present in everyone's subconscious.

It has been observed that over the years the West has easily adopted the tools and techniques developed at Toyota, because they are more visible, but has not so easily adopted the less visible aspects like routines, thinking and culture of Toyota. Figure 7.1 shows a representation of the differences between the more visible parts of Toyota and the less visible parts, which are routines, behavior and culture.

This fact is very easy to see by looking, for example, at the kind of books that have been published in the West about Toyota and the Toyota Production System over the last decades. The initial focus in the 1980s and 1990s was on the more visible parts of TPS, namely Just-In-Time, Kanban, 5S, SMED, Standard Work and Poka-Yoke. The human aspects more related to routines, teamwork, use of human capital and culture have only been looked at more

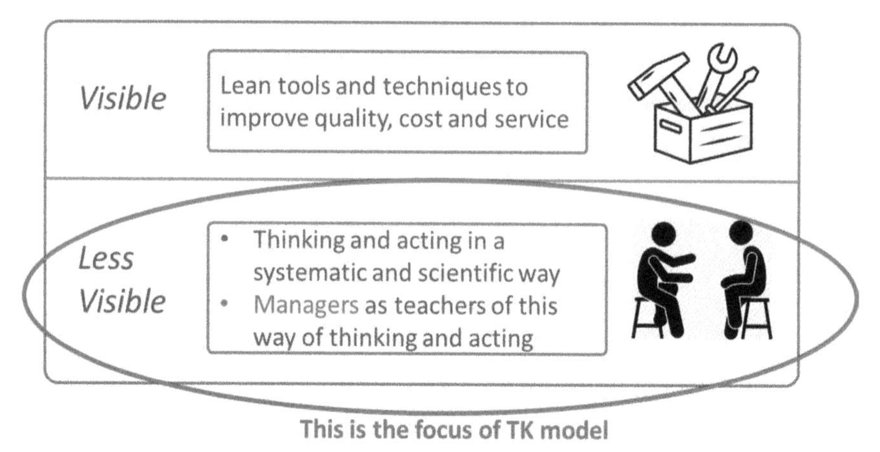

Figure 7.1 The visible and the less visible part of the TPS. Adapted from Rother (2020).

closely in the last decade. It is on this invisible part of continuous improvement that TK focuses on.

Scientific systematization is one of the important aspects of the TK model in order to make the continuous improvement effort more effective. This aspect of systematization is also evident in the "*Adopt Scientific Thinking*" principle of the Shingo Model. There is clearly an identification and appeal to scientific thinking and systematization as an effective and sustainable way of achieving results. These aspects of systematization and scientific thinking are central to Toyota's culture, with PDCA cycles and Standard Work as clear examples. The systematization and scientific approach of the TK methodology is represented in its four phases, which are presented in Figure 7.2.

According to this model, in phase 1 (Figure 7.2) the operational teams (*typically consisting of 3 to 8 workers*) should know the organization's vision (*true north*). The vision, defined by the organization's leaders, indicates the direction in which improvements should evolve. With this conscience of the direction where to go, the improvement actions to be implemented are prioritized over others. An example of a vision is, for example, to achieve "one-piece-flow" or "zero defects".

The size of a team is crucial, as it has an influence on performance. A good rule of thumb is to try to create small teams. On this subject, Sutherland (2014) states that teams of 3 to 7 people need 25% less effort than those of 9 to 12 people to perform the same work. Team size is often determined and/or forced by contextual constraints and by the personality and relationship typology among team members. There are teams of 3 elements that work very well, but others of the same size that do not work at all. The existence

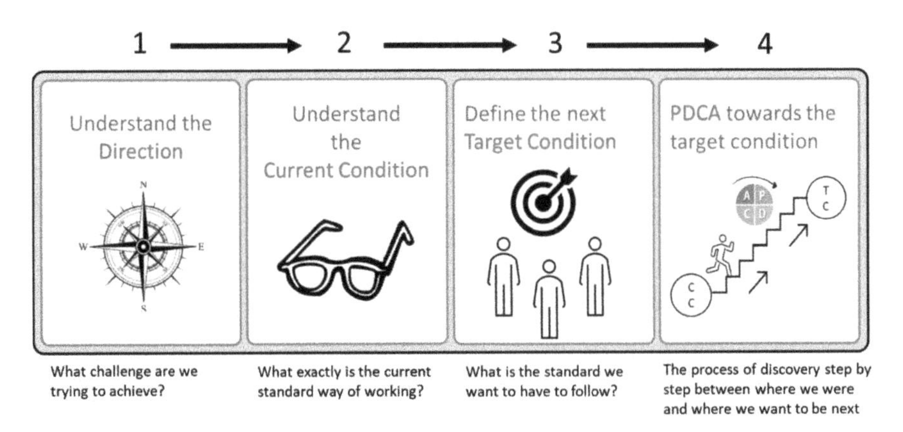

Figure 7.2 Phases of Kata methodology (Toivonen, 2015). Adapted from Rother (2020).

or not of a strong leader is also a key factor for the team's performance and for its ideal size.

In the second stage of this methodology (Figure 7.2), the team should have a good knowledge of the current situation (or current condition) in terms of performance indicators and form of operation (*layout, sequence of operations, operation times, work patterns, team skills, etc.),* presenting this information clearly and visually on the team board. In the next step, with the help of a Coach (the manager with a teacher role shown in Figure 7.1), the team establishes the target condition (*next target for the value of a performance indicator)* and the timeframe by which it commits to achieving it. The time frame (*period*) that the team commits to achieving this target condition is typically a few weeks. In the fourth phase, the team leader meets once a day with the Coach to evaluate the result of the last PDCA cycle (*last experiment*) and plan the next cycle (*which is typically one day*). When the team reaches the level of performance established as a target condition, the current situation is updated and a new target condition (*new performance level*), as well as a new deadline to achieve it, are established.

Regarding the fourth phase, where the team spends most of its time, a PDCA cycle is completed every day (Figure 7.3). The target condition should be assigned beyond the knowledge threshold. The team should not know how it would achieve that goal, it should accept it as a challenge outside its comfort zone, and that somehow, by experimenting with options it will get there with the help of the Coach.

The Coach has a critical role in the TK model: the main responsibilities are to maintain the Toyota culture, to keep the team motivated and to guide them to identify and implement improvement actions that guarantee learning, clear knowledge of the process and evolution towards the challenges and vision.

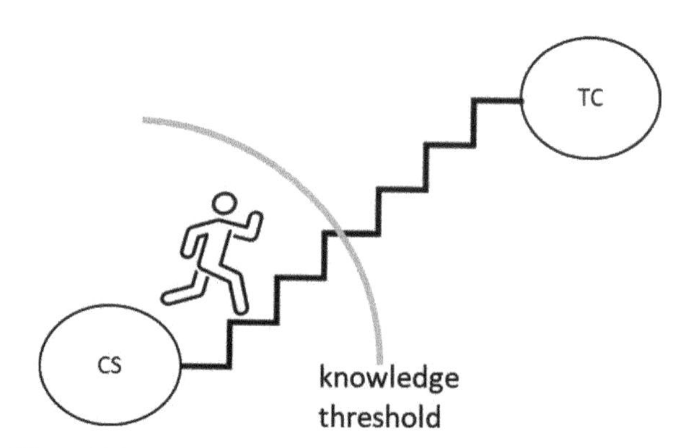

Figure 7.3 Target condition beyond the knowledge threshold.

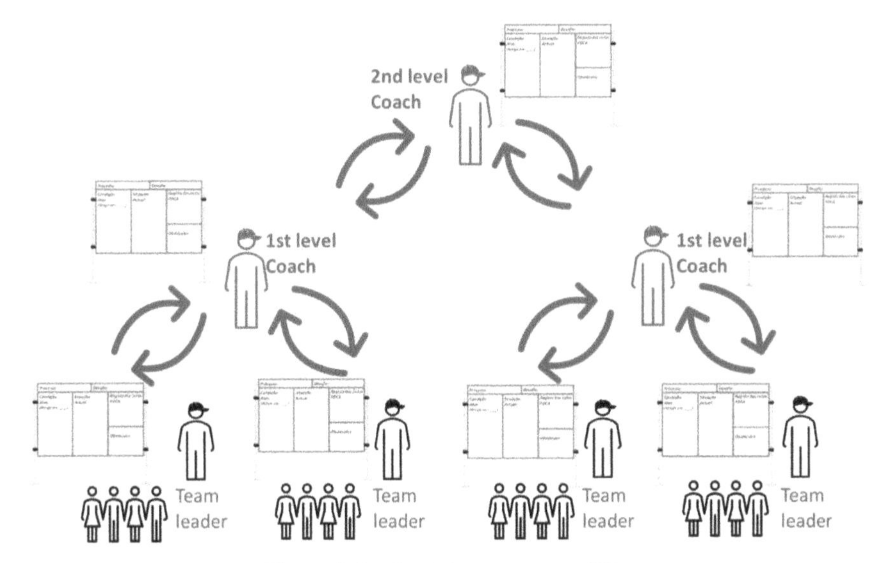

Figure 7.4 Coaching structure of TK.

The structure that guarantees the sustainability of the continuous improvement is represented in Figure 7.4. Each team leader meets with his coach once a day (whenever possible) for a session to follow the actions and their evolution always having as support the data that is visually presented in the team's chart. These daily sessions are occasionally accompanied by a 2nd level Coach that will help the 1st level Coach to evolve his coaching skills. When a vacancy for a 2nd level Coach opens, the 1st level Coach that is more advanced in terms of coaching skills will be selected. The vacancy left by the 1st level coach will have to be occupied by a team leader with more experience and skills in coaching. The worker of the team with more skills in leadership in turn will occupy this vacancy. In this way, the coaches are workers who absorbed Toyota's culture and take the responsibility to maintain it and pass it on to the new workers.

One of the main tools of the TK model is a team board that contains relevant information about the team and its work. The most relevant data to be displayed on the team board include, for example, the evolution of performance against defined KPIs, performance objectives, the next target condition, ongoing actions and a record of the PDCA cycles of ongoing experiences (right image in Figure 7.5). Based on all this information, effectively represented in terms of visual communication, the daily coaching sessions are carried out. In these daily sessions, the Coach always asks the team leader the same standardized sequence of questions as shown in the example summarized in the left part of Figure 7.5.

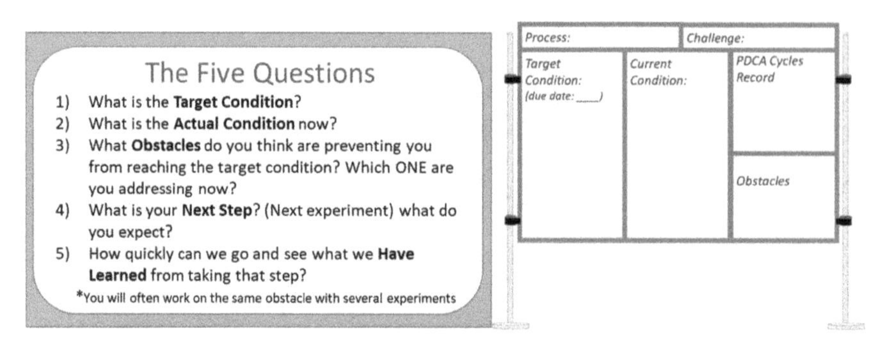

Figure 7.5 Coach routine questions (left) and example of a team board (right).

This standardized way of questioning is also perfectly aligned with the principles of creating routines, "kata", so associated with the development of unconscious automatisms (*the so-called "autopilot" mode*) that end up being very natural on a day-to-day basis. These actions and behaviors, being systematic and based on the scientific method, guarantee systematic and positive results of improvement in the medium and long term. This whole system of daily PDCA cycles of experiences, even when they do not result in improvements, generates learning about the process. The cycles of challenge, experience, result and feedback are extremely effective in maintaining the constant growth of people and improvements in processes.

7.2 KAIZEN Model™

The continuous improvement model named in this book as "KAIZEN Model™" and which will be presented here as one of the reference models is not published in its entirety in a book or in another publicly accessible document. However, some parts have been published in master thesis, in videos on Youtube, and some parts in books authored by the Kaizen Institute itself. In addition to this public information, I was also given access, by institutional courtesy, to internal documentation with more detail to better understand this model. Based on this model, which has been developed and continuously improved by the Kaizen Institute, that various instances of continuous improvement are designed and implemented in organizations around the world.

The Kaizen model assumes a structure with three pillars (Figure 7.6) covered by the big "hat" of Kaizen Transformation. These three pillars/dimensions have the following designations: Daily Kaizen; Leaders' Kaizen; and BreakThrough Kaizen.

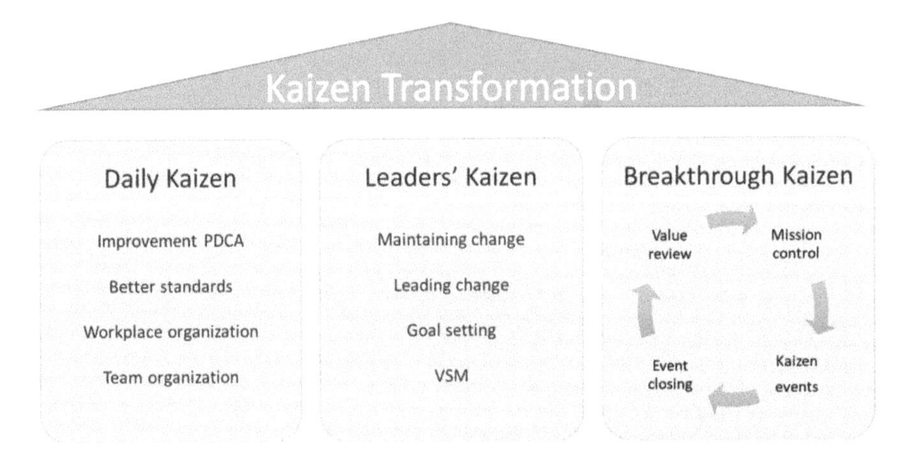

Figure 7.6 Kaizen™ Transformation model. Adapted from Kaizen_Institute, (2019).

Daily Kaizen refers to the operational aspects of the organization that include the people who carry out operations in the processes of transforming the products and/or services provided to the market. These are the people who directly add value to the products and services and who in practice are closest to the customer. They are closest to the customers in the sense that the product is produced or the service is provided by these people. There are also operational teams that, although they do not directly contribute to the added value of the products and/or services, contribute with internal support services to make the operation possible. Some organizations dub these services as "indirect areas". It is the daily kaizen that involves more people and is organized by teams that are usually called natural teams or operational teams as mentioned above. These teams typically hold a 5-minute meeting every day to be aware of the state of the team's performance, to be aware of the previous day's problems and the plan for the day. They are also expected to identify issues and opportunities for improvement and to be aware of the status of ongoing improvement actions.

These teams are responsible for ensuring the stabilization of the defined standards (Figure 7.7) and devote a small part of their time to continuous improvement. Daily Kaizen has the strategic objective of developing people and ensuring that the improvements introduced are maintained.

In the KAIZEN model™, natural teams go through four development phases until they reach maturity in the application of continuous improvement (Bastos & Sharman, 2018). In the first phase (level 1) there is a team board where team information is listed. The team holds a daily meeting (or as often as possible) for about 5 minutes in front of the board with information

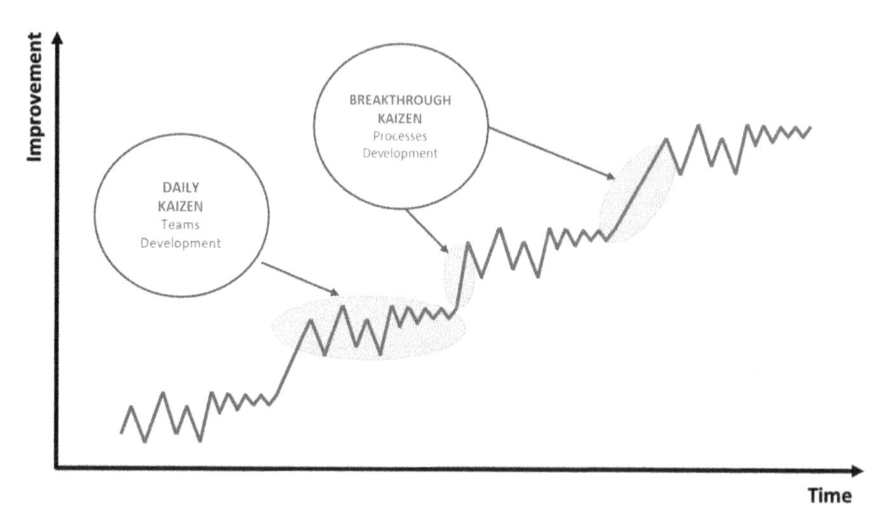

Figure 7.7 The continuous improvement process. Adapted from Kaizen_Institute, (2017).

support available on it. In this board are kept the updated indicators of the team's performance and the deviations are analyzed. These indicators are specific to each team but are typically related to the quantity produced, the quality and some type of efficiency and productivity.

In the second stage of maturity (level 2) the team also pays attention to the organization of the physical and digital spaces that are the responsibility of the team. The principle is that a clean and well-organized space is the basis for work to be done effectively, with quality and safely. In this phase, the team applies the 5S technique to make the workspace (physical and digital) organized and clean, where there is a place for everything and everything is in its place.

The next phase (level 3) is dedicated to the team's focus on standardizing tasks and processes. The creation of standards, for example the formal definition of how a task should be performed, is often undervalued because it is thought that people know how to do it or are already used to doing it in the right way. In relation to behavioral norms, it is also very common to assume that everyone implicitly knows what behavior should be adopted. The truth is that perception of this is a source of significant variation. Many times, different people do the task differently and with different results (the same reality exists in relation to behavior). In other cases, everyone resorts to a more experienced element to be the one to perform the task or to give instructions on it. The creation of standards involves codifying in a document the best (most efficient, safest and without ergonomic risks) way to perform a task. This document should use images as much as possible to make it easy and intuitive to follow and to help overcome language issues. Standards can also be used for the training of new team members.

The last phase of maturity of a team (level 4) is associated with process improvement, carried out by the natural team. At this stage of maturity, the natural team should also assume, according to Bastos & Sharman (2018), the responsibility of making small improvements in the processes and tasks of the team itself, and not just maintaining the standards as suggested in Figure 7.7, where the responsibility for process improvement lies with Disruptive Kaizen (it will be described below). According to Bastos & Sharman (2018), natural teams should use Toyota Kata as a mechanism for continuous improvement of their processes, although no detail is given on how the coaching structure in the Kaizen model takes place.

Another type of dimension with direct action on processes is called Breakthrough Kaizen, which is an evolution of the former Project Kaizen (this was the name used in previous versions of the same Kaizen model). Breakthrough Kaizen Teams are created to introduce disruptive process improvements in order to clearly improve the level of performance and typically lead to changes in working norms (standards). These teams are made up of people, who may be from different departments and with different skills, and they cease to exist when the project ends. The new standards developed by these teams are then followed by the natural teams to stabilize them as shown in Figure 7.7. These projects promote and introduce significant changes in the process that result in changes of layouts, change of tasks, resizing of lines, technological changes in the process, reduction of product change times, etc.

Breakthrough Kaizen is often materialized in Kaizen events that should typically be carried out for one week every 3 months. Kaizen events should start with a well-defined problem, described in a clear and short sentence, to enhance the effectiveness of the event and should be seen as an investment for its very effective role in employee training (McNichols, Hassinger, & Bapst, 1999). Bastos & Sharman suggest, that for there to be significant improvements in organizations, this level of performance called Breakthrough Kaizen, in conjunction with Daily Kaizen, has a multiplier effect in improving processes and consequently performance and competitiveness. The main message is that the combination of both types of efforts (Daily Kaizen and Disruptive Kaizen) for improvement result much more effectively than Daily Kaizen alone. However, Mike Rother presents a different view in his book Toyota Kata:

> *"... Toyota considers the improvement capability of all the people in an organization the "strength" of a company. From this perspective,*

> *then, it is better for an organization's adaptability, competitiveness, and survival to have a large group of people systematically, methodically, making many small steps of improvement every day rather than a small group doing periodic big projects and events.*"

The idea is that Toyota believes that an organization's greatest strength lies in the ability of everyone in the organization to make small improvements in a continuous and sustainable way. For that reason, for good adaptability, competitiveness and survival of an organization it is better to have a large group of people systematically, methodically, doing many small improvement steps every day, instead of a small group doing large projects and periodic events. Nevertheless, in the same book, the author also states the following:

> "... *If your business philosophy is to improve, then periodic improvement projects or kaizen workshops are okay but not enough...*"

Leaders' Kaizen is another pillar of the Kaizen model for sustaining continuous improvement whose relative position in this model is also represented in Figure 7.6 and Figure 7.8. While Daily Kaizen is aimed at all employees (direct and indirect areas), Leaders' Kaizen is aimed at all leaders/intermediate managers. The middle managers of an organization play a very important role in the success or failure of continuous improvement as they influence very directly, with instituted power, the attitude of the employees towards continuous improvement. Managers who are not sufficiently involved and motivated for the continuous improvement routines will certainly condition the employees in their attitude and contribution for this same continuous improvement. The first step in Leaders' Kaizen is then to train the leaders (managers) for continuous improvement with a training program within the context of the organization's own reality. The commitment of middle managers to continuous improvement is a key factor in the success of CI as well

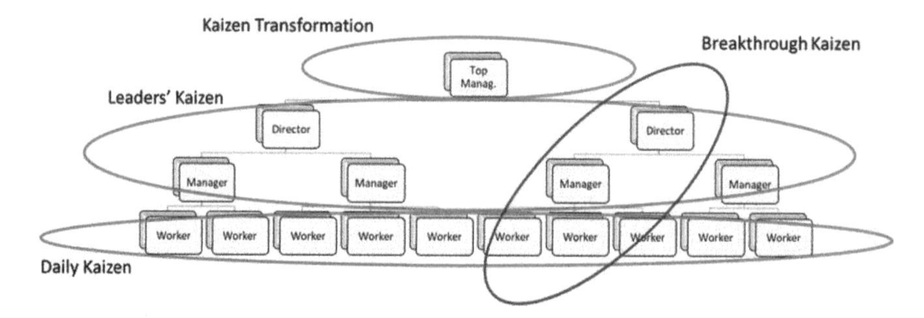

Figure 7.8 The several dimensions of the Kaizen model.

as its stability and sustainability in the medium term. Another very important role of middle managers is Policy Deployment, or passing on the strategic objectives and action plans to all levels of the organization, departments, sections and natural teams, so that all are aligned with the strategic vision of the organization, they are part of. Catchball is then used to systematically capture the feedback from the workforce and thus improve those objectives and plans. This creates the bottom-up energy and commitment emphasized by Devine and Bicheno in Creating Pull for Improvement (Devine & Bicheno, 2020).

In the Kaizen Institute model, managers go through a four-week training program to become active members and play a very important role in the continuous improvement effort. This program is carried out in the following five phases:

a. Phase 0: Planning - Raising awareness of the role of middle managers in continuous improvement. Furthermore, the macro planning for the implementation of the Kaizen model is elaborated;

b. Phase 1: Visual Management - Definition of the physical and/or virtual space that will support the visual management, with the objective of improving communication and initiating problem solving;

c. Phase 2: Commitment to the Gemba - Emphasis is placed on the importance of spending time on the ground in order to improve efficiency. Leaders should drive change by making frequent visits to workplaces;

d. Phase 3: Strategy Deployment and Implementation - Defining objectives and metrics aligned with the organization's strategy; and,

e. Phase 4: Strategy Improvement - At this level, it is intended that the strategies in force be challenged, with a view to improving their formulation.

Besides the already mentioned three pillars/dimensions, it remains to describe the top of the model; the level that sustains the whole system, the "Kaizen Transformation". In a way this level is the most important because it is at this level that it is firstly decided whether the CI movement will exist or not and it is also this level that defines and structures the three pillars of the system (*Daily Kaizen, Leaders' Kaizen and Breakthrough Kaizen).* Kaizen Transformation is the responsibility of the Transformation Leader who desirably is the Business Leader. To promote the supporting activities, a Kaizen Promotion Office (KPO) may be required. The Kaizen Transformation team should typically include the Business Leader and a group of people from the top of the organization who make up the Kaizen Promotion Office (KPO). KPO members should have expert knowledge and experience in continuous improvement, Kaizen/Lean principles, concepts and tools. This team has the role of assisting the natural teams on the ground as well as the Breakthrough Kaizen teams in terms of expert knowledge in continuous improvement, kaizen tools and defining the continuous improvement framework for the organization. The role of the Kaizen Transformation team also includes: training and coaching the organization's staff; providing regular and up-to-date feedback on performance, improvements and objectives; developing a continuous improvement culture; monitoring the teams and assessing results; and evaluating and auditing processes.

7.3 Scrum

Although the Scrum methodology is not, in essence, a continuous improvement practice, but rather an organization and management methodology for teams and projects, it is here presented because it assumes an organization structure with teamwork and includes in a very strong way a continuous improvement system. Furthermore, this methodology is very aligned with many of the concepts and principles outlined above and that we associate with excellence in organizations. Some examples are visual management, work visualization, constant PDCA cycles, fluidity of the work with limitation of the number of works in progress (WIP), teamwork, search for consensus, constant customer feedback, concern with people, leadership with humility, structured continuous improvement with "Sprint Retrospective", and others. Scrum is presented here also because it can be used effectively by teams that carry out the disruptive improvements so frequent in Lean environments, such as the aforementioned Kaizen events.

Regarding its origin, the term "Scrum" was initially used to describe a form of teamwork by analogy with the behavior of rugby teams at certain moments of the game, according to an article published by Takeuchi & Nonaka (1986) with the suggestive title "*The new product development game*". The authors of that article concluded that small self-organized and self-managed teams perform better in complex product development when goals are defined instead of tasks to be performed. The difference is that the solutions for the tasks and their execution work better when the team itself defines them according to the defined objectives. Executing a task knowing clearly what the purpose is, results better than simply executing a task. Knowing the "why" or "what for" of a task plays a very important role in the effectiveness and quality of that task.

The spirit of mutual help, perfect team unity and common purpose necessary for the effectiveness of Scrum training in rugby served as inspiration for how teams in other contexts could function successfully. The other contexts started as teams in product development in organizations such as Fuji-Xerox, Canon, Honda, NEC, Epson, Brother, 3M, Xerox and Hewlett-Packard (Takeuchi & Nonaka, 1986) and later, for example, in software development. An example of the use of similarly organized teams has been published by Coplien (1994) and refers to a Borland C++ software development project with impressive productivity. The Scrum approach has gradually been adopted with great success by the most advanced organizations in software development, according to Ken Schwaber in an article entitled "*SCRUM Development Process*" (Schwaber, 1997). In the same article, the following four methodologies used in product and/or software development are compared: the traditional or Waterfall methodology, the Spiral methodology, the Iterative methodology and the Scrum methodology. The traditional project management methodology, somewhat materialized in Gantt diagrams (*commonly referred to in the software development world as waterfall methodology),* is based on assumptions that often do not hold true in the context of software and product development. The difficulty in responding to unexpected occurrences in the various phases of the project is an example of the weaknesses of this methodology. The difficulty in "guessing" future tasks and their durations makes much of the traditional planning useless. This issue is quite well portrayed by Jeff Sutherland in his famous book on Scrum (Sutherland, 2014).

The Scrum methodology (Figure 7.9) is also inspired by Toyota's team management model (Sutherland, 2014) and the Toyota Production System itself, namely regarding the importance of flow, the concept of waste, the concept of customer value and the concept of continuous improvement or striving for perfection.

In Scrum methodology, each development team (*equivalent to an operational team or a natural team*) starts by meeting with the Product Owner (PO) to define the product/software features that need to be developed. The PO describes the product features in the form of User Stories. The set of stories forms the Product Backlog (Figure 7.9). In the context of software development, typically, a story is a description of a customer or user need that will be transformed into a small feature to be developed in the system. Some rules on how to write a story could be as follows: (1) the story should not contain implementation details; (2) the story should be focused on what the customer wants to do; (3) the story should be focused on the customer's point of view; (4) the story should be short and objective; (5) the story should contain the acceptance criteria. Very often stories are constructed starting with

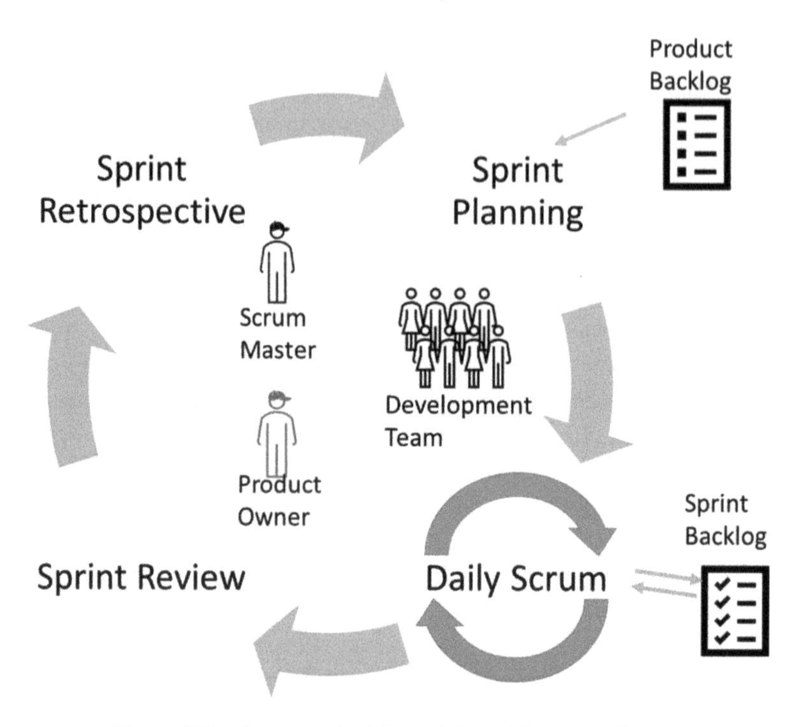

Figure 7.9 Scrum methodology. Adapted from Job (2019).

who wants what and why. More precisely each story starts with "how ...; or in the role of ..." followed by "I want ..."; "in order to ...". To give an example of a story we could have: "as a teacher I wanted an automatic system to register the entries in the classroom in order to assess the punctuality of the students". Note that no tasks are assigned to the team, it only describes, as best as possible, what the customer wants.

After all the stories are gathered, the Scrum Master meets with the development team and the PO to plan the first sprint (*Sprint Planning*). A sprint is equivalent to a PDCA cycle and typically lasts between 1 to 4 weeks. The team will do sprints until the end of the project so that in each sprint they learn from the best and worst results and improve performance throughout the project. The development team associates to each story a set of tasks and subsequently assigns an estimate in hours (or points) required to perform the task. To estimate the work required for each task, development teams usually use the so-called Planning poker (López-Martínez, Juárez-Ramírez, Ramírez-Noriega, Licea, & Navarro-Almanza, 2017) with the Fibonacci sequence. In the planning of each sprint, the PO defines the goal and priority of the sprint stories, while the development team adds stories, by the defined sequence, until it reaches its estimated capacity (*the velocity, in Scrum language*) of the team (*for example in hours.man*). During the product development phase, tasks are worked on according to their priority and are considered as finalized when they meet the requirements stipulated by what has been defined as "done". Every day there is a Daily meeting with the team and the Scrum Master to talk about what they have done, what they are going to do and if there are any impediments. At the end of each sprint there is a Sprint Review meeting with the PO and other Stakeholders in which the team shows the result of the tasks they did. In this way, it is possible to create a new version of the Sprint, which is then used as a reference for the development of a new version of the Sprint. According to Sutherland & Schwaber (2017) the purpose of the retrospective meeting is:

- Inspect how the last sprint went with respect to people, relationships, process and tools;
- Identify and rank order the major items that went well and potential improvements; and,
- Create a plan to implement actions/improvements concerning the way Scrum Team carries out its work.

This will be the last event of each sprint and the next event will be the planning of the next sprint. It is with this last event that continuous improvement is achieved (Sutherland, 2014) and it is for this reason that this methodology, although not always connoted as such, is presented in this book as an example of a continuous improvement model.

In addition to practices and routines, the Scrum methodology also uses some artefacts to support and enhance its effectiveness. Scrum artefacts provide key information to the team and other project stakeholders about the product, the plan, the status of different activities, and performance. The key examples of artefacts are: (1) Product Backlog - a list of features, functions, and requirements that the product should have. This list changes over time, as the product becomes clearer to the customer and to the development team itself and what is required to make it more effective; (2) Sprint Backlog - set of tasks that were selected from the Product Backlog by the team according to the priorities defined by the Product Owner to be part of the next sprint, and; (3) Burn-Down Chart - is a chart where the work that in theory needs to be done during the sprint is compared with the work that is actually being done. See an example in Figure 7.10 where in blue is represented the work that in theory should be finished in each day and in red the work that in reality was finished in each day.

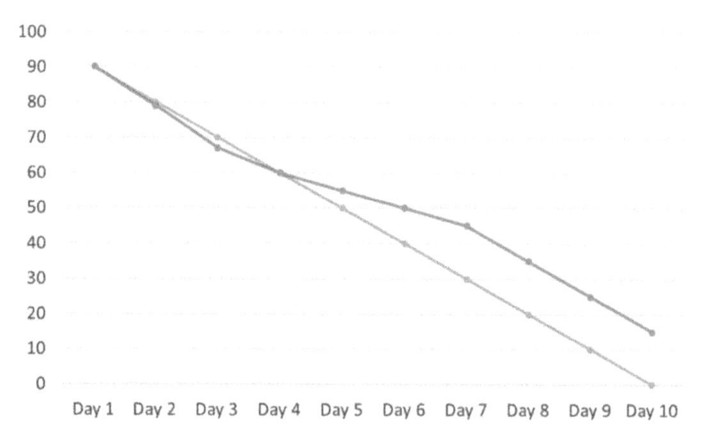

Figure 7.10 Example of a burn-down chart.

There is a very relevant aspect in the way the organization is structured for continuous improvement to work in a systematic, stable and sustainable way. The Product Owner has a role very close to the role of a boss in the hierarchy (*formal power*) while representing the customer. The Scrum Master has no hierarchical power and his role is very similar to the role of the Coach in the Toyota Kata model. The Scrum Master is an expert in the Scrum methodology and has competencies regarding organizing and managing teamwork and developing high performance teams.

One of the problems that Scrum teams face quite often is that the results of the retrospective meeting, in the form of suggestions for improvement, can be difficult to implement. This meeting is probably the most important event for improving team performance, since it is at this event that problems are brought up for discussion and improvement actions are proposed. It happens that if the team itself cannot carry out those improvement actions, they often are forgotten and are never solved. Some software companies have decided to have a development team only dedicated to carry out the improvement proposals that the operational teams of "developers" (*that is the term used in Portugal for the members of the software development teams).*

7.4 Comparison of Models

One can start by saying that a key point that is common to the three models presented for systematizing CI is that they are all based on practices identified at Toyota. The Toyota Kata model is presented by Rother (2010) as being obtained by interpreting routines observed by him at Toyota. The Scrum model is also clearly presented by one of the main authors (Sutherland, 2014) as having been based on practices observed or inspired from Toyota. Regarding the Kaizen model, it is a model that although being inspired by Toyota's practices it is not clear, at least to us, that it was developed according to Toyota's practices regarding the systematization of continuous improvement.

Another common and central feature in the models presented is that they all assume an CI structure in which employees are organized into teams. We call these teams, operational teams although they are also often referred to as "natural teams", especially in Kaizen model implementations. These operational teams are the ones who are closest to the value-adding operations for the products and/or services provided to customers. This aspect is very relevant because, as on the one hand, it ensures the involvement of these employees in continuous improvement and with it a sustained improvement in the organization's competitiveness, and on the other it represents a gain for the employees by providing them with a favorable platform for personal development and job satisfaction. There is, however, a difference in the models presented in what is expected of the role of operational teams. In TK it is expected that these teams contribute with the introduction of small improvements every day, which is the basis to guarantee the sustainability and stability of the continuous improvement. In the other models, this is different. Although the teams hold daily meetings, in the Kaizen model the main objective is that the team stabilizes the work standards leaving to the kaizen events the responsibilities for the more significant and disruptive improvements. In relation to the Scrum model, the definition of improvement actions by the natural teams (*in this case, development teams*) is expected to happen in a systematic way at the end of each sprint (*between 1 and 4 weeks*) in an event called Sprint Retrospective. In fact, if we make an analogy with the PDCA cycles, at the end of each sprint, this Scrum phase called "Sprint Retrospective", would be the equivalent to the "Act" phase of the PDCA cycles. Using the same analogy, the "Sprint Review" phase of Scrum would be equivalent to the "Control" phase of the PDCA cycles.

Many organizations using Scrum as a working methodology in their operational teams, typically software development teams, or also development teams of other products, manifest some difficulty in implementing the suggestions that are generated during these retrospective phases. This is a very common problem, especially in organizations that have not yet reached advanced stages of maturity. The generation of improvement ideas or problem identification happens in many organizations at a much higher rate than the capacity to implement those suggestions. In more mature organizations in terms of continuous improvement, as is the case of the software development *Company_P*, a team of 10 engineers is attributed the task of developing and implementing the validated suggestions. This was the way the company found to avoid the suggestions to be forgotten with a double negative effect. The first negative effect is that opportunities for improvement are lost and the second negative effect is to create frustration and loss of commitment in the operational teams.

The Scrum methodology was developed specifically to organize and manage the work of software development teams although it is already being used effectively in development teams of other types of products. It is important to note that this methodology is not suitable for all types of work that are carried out in teams. The types of work for which this methodology is suited include:

- Diverse tasks in a specific knowledge area that require intellectual work and some creativity;
- Types of tasks on which there is experience in the team;
- Types of tasks that may require from a few hours up to days of work; and,
- Types of tasks for which you can estimate duration times with some slack.

According to these characteristics, it is clear that this methodology is not suited to teams performing repetitive tasks of short duration (*seconds or minutes*) which are carried out for example by teams working on assembly lines or in industrial production in general. In this type of teams, Toyota Kata or Kaizen methodologies are much better suited. At the other extreme of the typology of team tasks requiring intellectual work that demands creativity and/or scientific work, not even the Scrum methodology can be applied effectively. In this type of work, where the task duration is extremely difficult to estimate, it is more appropriate to use *kanban* boards with limited WIP.

The *kanban* board technique oriented towards the management of teamwork or individual work, very often used in intellectual work environments (*in particular research and development*) is not referred to here as being equivalent to the three methodologies presented. The reason for this is that this methodology does not include specific characteristics of continuous improvement. It is only referred here to clarify the differences and similarities that exist between it and Scrum. More information about *kanban* boards can be found on websites or in books such as "*Personal Kanban: Mapping Work | Navigating Life*" by Jim Benson (Benson, 2011). It is perhaps worth mentioning that *kanban* boards are inspired by the *kanban* systems used to control the progress of materials in production flows in industry but are

Table 7.1 Comparison of different methodologies for teamwork.

Methodology	Kaizen	Toyota Kata	Scrum	Kanban
Type of tasks	Repetitive and undemanding	Repetitive and undemanding	Varied but predictable	Varied and unpredictable
Tasks duration	Seconds or minutes	Seconds or minutes	Hours or days	Days or weeks
Monitoring mechanism for teams	Audits	Coach	Scrum master	Not structured
Examples	Industrial production teams	Industrial production teams	Development teams	Research teams

actually quite different uses and practices. What these two uses have in common is the use of cards and the limitation of WIP, which in the case of *kanban* boards used in teams is much less accurate.

Table 7.1 contains a summary comparing the three methodologies presented with characteristics of continuous improvement and the *kanban* technique to support the management of individual or teamwork.

The Scrum methodology, even though it is used intensively in the software industry, stops being used when a team needs to find out the causes of errors in the developed software. In these cases, teams stop using Scrum and adopt *kanban* boards instead. The reason for this happening is due to the great unpredictability of the times required for the duration of software error resolution tasks. Whenever there is such unpredictability Scrum is not a good methodology and *kanban* boards serve this purpose better. It is important to note that some of the benefits of the Scrum methodology are not achieved with the use of *kanban* boards. One of the examples is the greater structuring of the methodology, with the systematization of sprints, the PDCA nature and the continuous improvement achieved with the "Sprint Retrospective" phase.

Regarding the team coaching mechanism, the Scrum and Toyota Kata methodologies present an equivalent mechanism. There is a person with coaching competencies, who knows the methodology and has teamwork skills. In Toyota Kata this person is called Coach and in Scrum is called Scrum Master. This person maintains a daily follow-up of the team, helping it to grow and constantly improve its performance. The concern with team spirit and the state of interpersonal relationships among members is much clearer in the Scrum methodology. Still regarding the team follow-up mechanism, in the Kaizen methodology this function is ensured by the kaizen team, which, in audits, detects the team's needs and problems and acts accordingly with appropriate improvement actions. Normally, the focus is not on people but more on processes. As for the *kanban* boards, there are no clearly foreseen mechanisms for team monitoring.

The question we may ask here is what type of model should an SME company adopt? What will be the most effective model for organizations with dozens or a few hundred employees? This question will have several answers but, in the next chapter, we will try to give some suggestions on possible alternatives.

References

Bastos, A., & Sharman, C. (2018). *Strat to Action - O Método KAIZEN™ de levar a Estratégia à Prática.* Kaizen Institute.

Benson, J. (2011). *Personal Kanban: Mapping Work | Navigating Life.* Modus Cooperandi Press.

Coplien, J. O. (1994). Borland software craftsmanship: A new look at process, quality and productivity. *Proceedings of the 5th Annual Borland International Conference, Orlando, Florida.*

Devine, F., & Bicheno, J. (2020). Creating Employee 'Pull' for Improvement: Rapid, Mass Engagement for Sustained Lean. In *Lecture Notes in Networks and Systems* (Vol. 122). https://doi.org/10.1007/978-3-030-41429-0_7

Job, J. (2019). *My Scrum Diagram.* Available at https://jordanjob.me/blog/scrum-diagram/. Accessed at 14/04/2021.

Kaizen_Institute. (2017). *Kaizen Internal Report.*

Kaizen Institute. (2019). *Kaizen Institute.* Available at https://mt.kaizen.com/. Accessed at 19/03/2018.

López-Martínez, J., Juárez-Ramírez, R., Ramírez-Noriega, A., Licea, G., & Navarro-Almanza, R. (2017). Estimating user stories' complexity and importance in scrum with Bayesian networks. In *Advances in Intelligent Systems and Computing.* https://doi.org/10.1007/978-3-319-56535-4_21

McNichols, T., Hassinger, R., & Bapst, G. W. (1999). Quick and continuous improvement through kaizen blitz. *Hospital Materiel Management Quarterly, 20*(4), 1–7.

Rother, M. (2010). *Toyota KATA: Managing People for Improvement, Adaptiveness and Superior Results.* McGraw-Hill Education - Europe.

Rother, M. (2020). Toyota Kata website. Retrieved March 22, 2018, from http://www-personal.umich.edu/~mrother/Homepage.html.

Schwaber, K. (1997). SCRUM Development Process. In *Business Object Design and Implementation.* https://doi.org/10.1007/978-1-4471-0947-1_11

Sutherland, J. (2014). *Scrum: The Art of Doing Twice the Work in Half the Time*. Penguin Random House.

Sutherland, J., & Schwaber, K. (2017). The Scrum Guide, the Definitive Guide to scrum: The Rules of the Game. Retrieved from https://www.scrumguides.org/scrum-guide.html.

Takeuchi, H., & Nonaka, I. (1986). The new product development game. *Journal of Product Innovation Management*. https://doi.org/10.1016/0737-6782(86)90053-6.

Toivonen, T. (2015). Continuous innovation - Combining Toyota Kata and TRIZ for sustained innovation. In *Procedia Engineering* (Vol. 131). https://doi.org/10.1016/j.proeng.2015.12.408.

8

Developing CI in your SME

This chapter is intended to help the reader to lead or to help initiate the process of developing a culture of continuous improvement in his/her company or organization, but especially if it is an SME. There are many possible paths and it will be difficult to guarantee which will be the best one, but here some tips are given that can be valuable to prevent the difficult missteps that are too often made. There will be many other ways to achieve good long term results but we believe that these tips presented here can avoid some frustrations. Of course, the reader should be prepared for some frustrations along the way but that is the price you have to be prepared to pay for doing something that is really worth doing. In this case it is worth it because it has the potential to improve your organization's competitiveness, the satisfaction of employees and managers, and contribute to improving the community and the local economy. The tips presented here are the result of many experiences carried out in several SMEs in the north of Portugal, and from these experiences resulted many learning moments either by success or failure.

8.1 Where to Start

Let's assume that you want to help an SME to start or breathe new life into continuous improvement. It can be in the SME where you work or in an SME where you will work in the future. Let us try to indicate what would be the best way to start. First of all, a necessary condition is that you have read all the previous chapters of this book. Besides having read them it is important that most of what you have read has made sense to you and that you unreservedly review the principles presented by the main TPS-inspired excellence models for organizations.

In accordance with what Masaaki Imai (founder of the *Kaizen Institute*) likes to say, it is necessary to understand that there are 3 essential conditions for the successful implementation of continuous improvement. These 3 conditions are: firstly management commitment, secondly management commitment and finally thirdly management commitment (now the reader should remember what was said in chapter 4 about the true meaning of commitment when we referred to the role of the pig and the hen in the English breakfast). Instead of management we can also say 'the number one' in the organization, as Bastos & Sharman (2018) refer to, i.e. the person who has the final word in decision making. If you can get that person to read a book like this one you are reading, who is willing to listen to you about what you already know about continuous improvement and excellence in organizations, or willing to take a short course on CI, *Lean* or *Kaizen*, then it may be that this decision-maker is truly willing to accompany you to carry out the necessary transformations. The necessary conditions for this "number one", who is often the owner of the company, to be your great asset for transformation is necessary two conditions: (1) that this "number one" fully understands the direction in which the organization is heading and what needs to be done to move in that direction, and (2) that this "number one" believes in people.

Now, if the person who has the greatest power of decision does not meet the conditions presented above, then it is not easy at all, but that does not mean that it is not possible. You can follow another path, much more difficult and slower, which is to keep trying and keep showing that person the power of the CI. Remember that we are all here to learn and grow and so this person can also learn and grow. In the meantime you can move from the bottom up with CI initiatives in a production unit or some service in the organization where people are willing to work with you. Try to create a small team that supports you and is with you in this journey because if you are alone it is almost impossible. With time and with the results you achieve, you may be able to get other units or departments of the organization to follow your example and when the results appear the leader of the organization may start to get curious about the process. Starting from the bottom up is not easy and obviously depends on many variables, but although hard, it is a possible path.

8.2 Don't Be Afraid to Fail or Do It Differently

Do not assume that solutions that work in other organizations will have to work in yours. Of course you should learn from examples of solutions from other organizations and you should know as much as possible what other organizations do, but don't just copy. The reality of the organization you work for is very particular and there is no other organization like it. Its products are different, its processes are different, its people are different, its culture is different, in short, it is different from all the others. Since it is different from all the others then its continuous improvement system, routines, standards and artefacts can naturally be different too. The only thing that can be the same are the principles of excellence in organizations that were presented in chapter 2 and the concepts that have been presented throughout the book. These principles and concepts should serve as a guide for the idealization and selection of improvement actions and "solutions".

The word "solutions" appears in inverted commas because in truth one never gets "the solution". One can never reach "the solution" because there will always be room to improve what one already has. In fact, what we have are always intermediate or temporary solutions to improve something. Given this clarification, the word "solution" will continue to be used although, as we have seen, it is not precise in its meaning.

Returning to the issue that solutions must be designed for each case, the example of system solutions and mechanisms to manage material flows is a very paradigmatic example of this. If the dear reader enters several organizations that have flow management systems such as *kanban* systems, supermarkets, *mizusumashis*, etc., you will agree that all systems adopted are different. Although all are guided by the principles of the pull flow production and are inspired by other cases, the truth is that in each case there is a solution specifically designed to suit the characteristics of that case. The mechanisms you design to ensure flow should best serve the characteristics of your processes. Don't be afraid to experiment with innovative mechanisms to suit your specific context. Try and see the result in a spirit of PDCA cycles. In each experience there will be new learning and evolution.

Regarding finding solutions for each context here is an interesting example. A small operational team from a company named MoldartPovoa identified as one of the main problems the long setup times in their main equipment, which was around 2 hours. One of the team's indicators to be constantly monitored and improve was the setup time of that equipment. As some of you will know, the classic technique to reduce setup times is the *Single Minute Exchange of Die* technique (see Appendix A). The classic implementation of this technique usually requires assigning this

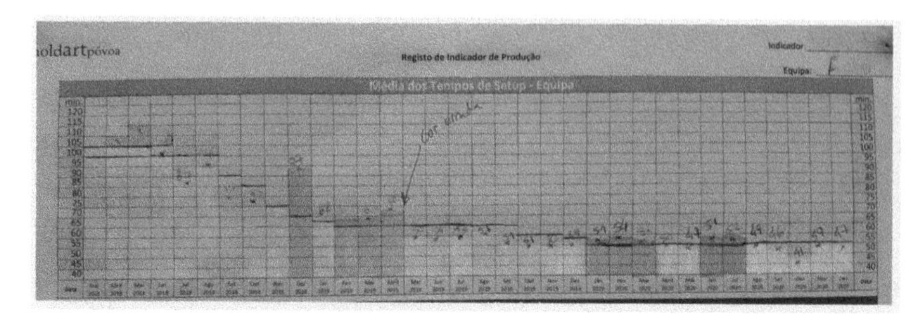

Figure 8.1 Evolution of setup time in a team's machine.

work to a person or a team to work on it for some time. As there were no people in the company with experience in that technique and who could dedicate time to this subject, the adopted approach was quite different. It was decided to assign the operational team the task of finding small solutions, in a CI spirit, to reduce the setup time of the equipment.

The team introduced small changes in the way they prepared the machine and we can share some examples as follows: They started to do some setup tasks while the machine was still running that they normally did with the machine already stopped, they changed some tightening systems for solutions with faster tightening, some tasks that they used to do in a sequential way started to be done in parallel, they changed the sequence of tasks to reduce the movements, they moved closer the parking places of tools that they needed during the preparation of the machine, they improved the positions of tools and other devices using 5S technique, as well as other solutions. In this way, over time they reduced the setup time from around 2 hours to around 50 minutes as we can see in the record shown in Figure 8.1.

This is a perfect example of how effective solutions can be found through small improvements without resorting to techniques and solutions that are normally used or created by others. The important thing was that the team accepted the challenge, understood the problem, closely monitored that indicator and continually made small experiments and changes to their procedures. In each experiment they checked the impact on performance and each time they improved performance they took on a more ambitious goal. They were not afraid to fail.

The artefacts, such as team boards, must be adapted to team's reality, their restrictions and their potential. And most important of all, don't be afraid to make mistakes and don't be afraid to change or give up on solutions that initially seemed to be the best ones. Get this message across that it is important to try and experiment, and that it's okay that what we thought would work doesn't work after all. Whenever you try something that does not work, there is a positive part of learning. Sometimes even the failed attempt is the source of inspiration or the way to a great and completely innovative solution.

8.3 Continuous Improvement Management Team

If you have already made the decision to start on the path of continuous improvement you should start by creating a CI management team. This team must be formed by you and two or three colleagues who understand and believe in continuous improvement and in the importance of pursuing organizational excellence so that the organization they work for ensures sustainable growth. It is very important that a person from the administration board or top management belongs to this team. We believe it might be worth it to hire an experienced CI consultant to help you with their knowledge and experience with CI in other organizations. This consultant can be considered as part of your CI team. If they choose not to hire a consultant then at least one of the team members must already have knowledge and some experience with CI, *Lean, Kaizen,* and/or any other model of Excellence in organizations inspired by TPS.

There is a classic mistake that some decision-makers make regarding how they believe transformation should occur in the organizations they lead. That mistake is to invest large amounts of resources with a large firm of consultants in order for that firm to bring about a rapid transformation to continuous improvement and excellence in the organization. This is another mistake that goes against some of the principles of excellence in organizations. I am referring to the principles of the long-term philosophy and the principle of pull flow production. I hope the reader remembers what was talked about earlier about the virtues of piece-by-piece production rather than batch production. The transformation to continuous improvement should also not be done all at once (like a big batch), it should be done step by step, learning and consolidating at each step. But don't misunderstand, it's not advocated that you should avoid large consultant firms, it's just advocated that you should do it slowly over a long time, investing little at a time, rather than quickly with a lot of investment. The transformation process should be aligned with the speed of learning and you cannot burn steps. There are some examples of serious organizations that made a big investment to quickly implement continuous improvement and excellence practices in their organizations and a year later nothing was accomplished. I know one organization that one year after the consultants completed the project, the team boards were no longer updated, floor marks were ignored, pull systems were replaced by verbal orders from supervisors, daily meetings were no longer held, performance indicators were no longer updated, and so on.

The CI management team should start by applying to itself what they wants other teams to do. This team should have a team board with monitoring of their performance indicators, standard agendas for meetings, plans for actions, monitoring of ongoing actions, and other information they find important. In addition, they should hold regular meetings and maintain routines for the operational teams. It is important that your CI management team members practice to the best of their ability the principles of excellence in organizations that were presented in chapter 2. There are several principles that need to be discussed and clarified but here probably would be elected the **continuous flow production** as one of the main principles and paradigms to be clearly understood by all. If you think it is good you should do some short course on continuous flow production so that everyone in the team understands and lives this principle. If any of the team members have doubts about the benefits of this kind of approach to production then they should not continue until that subject is perfectly understood by all the team members. This should be said because otherwise there may be conflict in the way messages are passed on to employees and this leads to difficulties in deciding the improvement actions to be implemented on the ground. Do not underestimate the importance of everyone's alignment in the pursuit of continuous flow and production pulled by demand.

An important attitude that should be avoid at all costs is following: members of the CI management team cannot be spreading principles and practices to the operational teams and in general throughout the entire organization and then not following those same principles and practices on their day-to-day basis.

On this subject of "practice what you preach" I'm going to tell you a story of a CI leader I met in a partner organization and with whom I ended up forming friendships. He is an excellent speaker and excellent manager and motivator of people and was very successful in transforming the organization he worked for, which was going through bad times almost two decades ago, into an organization with great competitiveness in their field of business. The transformation achieved with the implementation of Lean concepts and tools was so effective that in addition to improving all of their productivity, quality and service indicators, in relation to the products they already produced, they also managed to attract to their production many new products that were not in their core business. The first time I entered this friend's office at the organization, I was amazed by the mess, both on his desk and in the entire office space. I was so amazed that I couldn't avoid asking him how that could be possible. He responded with a famous proverb (at least is famous in Portugal) saying something like "...You know, In blacksmith's house, wood skewer". But he said it with a certain guilty look on his face. He is my friend and I have great respect for him but this kind of misalignment must be avoided at all costs. Employees cannot be told to follow what leads say when those leaders do not act accordingly. It is more real the following saying "Don't listen to what people say, watch what they do". I hope you agree with me.

About the continuous improvement management team, we can admit that ideally it should have as one of its purposes the end of its existence. The idea is that continuous improvement gradually becomes part of the work of everybody in all levels of the hierarchy. When continuous improvement is completely integrated in the organization's strategy then the whole organization, by strategy or policy deployment, only has to align its objectives and its work with the organization's strategy, as we saw with the use of the *Hoshin Kanri* methodology. This way it is possible that in an ideal organization there is no specific team or people to worry about continuous improvement since it (the CI) is an integral part of the organization's nature and culture. Nevertheless, until we get there it is good to have a team that is preparing the organization to get there.

What we have seen in all organizations that seek continuous improvement and excellence is that in addition to having a continuous improvement team, there is also a "number one" for the continuous improvement function, just as there is a "number one" in every company or organization. It is the quality and vision of this "number one" that in many organizations is the key for the survival of the CI and as a consequence of the organization's long-term success. This "number one" of the CI team, which is probably you, has to gather some very important characteristics where we can include at least: Resilience, persistence, patience, CI knowledge, leadership skills and ability to work in a team.

There is an aspect that is not commonly mentioned in the literature, as it is assumed to be intrinsic, or taken for granted, but which plays a very relevant role in the success and effectiveness of CI. It is about the existence of someone who is dedicated to implementing small solutions that were suggested by the operational teams and then accepted to be implemented. For large actions and projects, the CI management team usually assigns a specially created team to carry out this implementation, as is the example of project teams. However, the many small improvement actions may remain to be implemented because there is no one with that role within the organization. These small actions include examples such as drilling holes in the wall to hold things together, making small devices out of wood or metal, painting floors and walls, gluing tapes, developing and printing signs, and other types of "odd jobs". There is always some worker who has the art to carry out this type of work and assigning him that role can be an interesting solution. The existence of a "handyman" dedicated to responding to constant requests to implement small improvements can be a good solution. This person does not have to be 100% of their time dedicated to these tasks but it is great to be always available at times when necessary. There are organizations that assign these small actions to the maintenance team, but it does not always work very well because these teams can take a long time to be available to carry out these implementations.

There are several organizations having difficulties in effectively implementing all the valid suggestions presented by the operational teams. In organizations where the Scrum or DevOps model is applied, and they are not just organizations with production in the IT area, the operational teams in the retrospective phase of each sprint can generate a lot of suggestions for improvement that often remain to be implemented. These suggestions remain to be implemented because there may not be resources in the organization dedicated to implementing these improvements. It is the lack of that one or more "handyman" figure referred to in the previous paragraph. One way to solve this problem is to believe that it is worth assigning human resources to tasks related to the implementation of all validated suggestions. This can be very worthwhile indeed. An example of this is Company_P presented in chapter 0, which created a team of more than 10 engineers whose sole function is to implement the suggestions of the operational teams in the field. Despite being a huge cost, the assessment that Company_P has been carrying out is that the benefit in productivity gains, speed of response and quality, are much greater than the costs of that implementation team.

8.4 Teamwork

Teamwork seems to be unquestionable in continuous improvement contexts. The importance of teamwork on the path to organizational excellence is always present in the models used as a reference in this book. This importance is clearly expressed, for example, in the tenth principle of the *Toyota Way* that says "*Develop exceptional people and teams that follow your company's philosophy*". Although there is no principle in the Shingo model or the *Lean* Philosophy that clearly expresses the issue of teamwork, its existence is expressed in the descriptions of the models and there are several references to its importance. Already in the first books on the *Toyota Production System*, the theme of teams and teamwork was present in the main concepts of that system. You can find references in classic books on TPS such as the original book by Taiichi Ohno where he introduces TPS to the world (Ohno, 1988), but also in the famous book by Yasuhiro Monden (Monden, 1998) and many others. Taiichi Ohno even attributes the suggestive title "*Teamwork is Everithing*" to a section of his book. The author states that in Japan the tradition is individual sports such as Judo and Karate while team sports were more traditional in the West and therefore they had to learn the power of teamwork. To clarify the importance of teamwork, the author uses the example of rowing using the case of four rowers on each side. In that case, it would not matter if one of the rowers tried harder than the others as this would result in an erratic boat trajectory. The most important thing is that everyone paddles to the rhythm of the team and as a focus on the whole in order to win. In terms of conclusion, it is important to be aware that teamwork plays a fundamental role in the origin of models of excellence in organizations and consequently in continuous improvement.

The importance of teamwork is also frequently mentioned in books that presented the Lean philosophy. This is clear both in the famous "*The Machine That Changed the World*" by James Womack, Daniel Jones and Daniel Ross (Womack, Jones, & Roos, 1990) and the also famous "*Lean Thinking*" (Womack & Jones, 1996) where the 5 *Lean* principles are presented and, in addition, the importance of teamwork in environments where the *Lean* philosophy is pursued is reinforced.

That said, if the reader is in a position to start taking steps towards creating a continuous improvement system or movement in your organization, you will have to think about creating teams. The idea is not just to create operational teams; the idea is to create teams across the entire structure. Start by creating the CI management team where you and two or three other colleagues who believe in continuous improvement and excellence in organizations. Then create one or two operational teams that serve to put teamwork into practice and learn from the experience of your reality. Do not be tempted to start with too many teams because then you will not have the energy to sustainably follow up all the work. The result is that, if that happens, the teams will feel unaccompanied, they will lose enthusiasm and when they want to start again it will be more difficult because many people have certainly started to disbelieve that path. Start slowly by taking small steps at a time. Remember that creating teams is not just saying that that group of people becomes a team. It is not that easy. It takes a lot of work for groups of employees to become real teams that thinks like teams with focus on common goals. If you manage to create motivated teams that live the team spirit, then you already have a vital platform for success in the continuous improvement and excellence of your organization.

To create an operational team it is necessary to dedicate time to that team. You will need team members to believe in you and they will need to feel that you are genuinely there to help them improve their well-being and job satisfaction, while producing more, faster and with more quality. If you meet with them frequently using for example the *Toyota Kata* techniques that were presented in chapter 7 they will gradually be on your side and will be able to do fantastic things. Afterwards, it is important that the team gain more and more autonomy and responsibility, assuming for it, as much as possible for the context in question, a large part of the management and organization of the work. The more management and continuous improvement tasks are transferred to the team's responsibility, the greater the availability of middle managers to deal with innovation (disruptive improvements) of their products and processes. The involvement of everyone in management is a key aspect of Toyota's success as it develops a culture where everyone works in harmony for the same cause and this generates great personal satisfaction. This aspect is referred to in the book "*40 years, 20 million ideas*" by Yuso Yasuda (Yasuda, 1991).

8.5 Operational Teams

Operational teams are all the teams responsible for the operation in both the direct and indirect areas. To clarify this difference, considering an industrial company, we will have operational teams responsible for transforming raw materials into final products (cutting, welding, painting, etc.) but also operational teams in indirect areas such as planning, logistics, accounting, purchasing, commercial, product development, etc. This second type of operational team is not directly responsible for adding value to the products the company provides to customers but for carrying out tasks in areas that are critical for the organization to function and keep it running. The formal existence of operational teams, whether of one or both types, is common to all CI systems we know and is probably the element whose existence seems to be unquestionable.

Operational teams are organized and work differently from organization to organization. In terms of size, for example, the variations observed are enormous. At the Ikea factory in *Paços de Ferreira (Portugal)*, for example, operational teams typically vary in size between 5 and 7 members, one of whom assumes the role of team leader and wears a t-shirt of a different color (orange color)) to be easy to identify. At the Gewiss factory located also in our region, in *Penafiel*, the size of the teams is not so standard. Each production cell constitutes an operational team and its size varies between 4 and 20 elements. Each team has been assigned a team leader although more than one team can share the same leader. It is debatable whether two teams can have the same leader, but this was the solution found by the organization to possibly overcome some restrictions or constraints. It is important to accept that successful solutions from one organization may not be successful at any other organization. The important thing is for each one to find the solutions that best serve the objectives in the context of the organization in order to make the organization evolve in the direction defined in its vision.

Teams that are the totality of operators working in a production unit, even if it is large, was also observed in Company_B (a large company in the metal working sector). In this organization's factory, the size of the teams varies between 9 and 30 people. Each team is associated with a leader who, if the team is large, can have the support of an assistant. The effectiveness and efficiency of teams with such large dimensions is very debatable as previously stated. In teams of 30 elements, some difficulty in developing team spirit can be expected. It is hard to manage team meetings in an effective way in such big teams. Meetings with such large teams often comes down to a presentation by the leader conveying information regarding performance, plans and recommendations without effective interaction between team members. Many of the team members of these dimensions assume a passive position and often even oblivious towards the presented subjects.

This problem of large teams is also observed at *Lipor* in *Ermesinde* (Portugal). The operational teams in indirect areas, that is, the support departments (more administrative and intellectual operations) vary between 3 and 14 elements, but in the departments where the actual physical production tanks place the teams vary between 5 and 97 elements. The organization decided that the entire infrastructure and logistics division with 97 people working in three shifts would be just one team. Here we have the issue of the size of the team, which is too big, and then there is also the issue that the team is divided into 3 shifts. Elements from one shift may not even know elements from other shifts. This was, however, the solution that the organization found within the restrictions, limitations, culture, labor relations and context of the moment that best served the purpose. We can assume that with the learning achieved with this solution they can evolve to effectively divide this large team into smaller teams with better results in the future.

For you reader, our advice is to create teams with a number of elements between 3 and 7 but in some cases it may be acceptable to create teams a little bigger. We advise against creating very large teams because their effectiveness can be harmed. If you try to take the principles of teamwork to a very large team, it will not have great results and therefore we advise you to divide these large teams into smaller teams with their own goals and routines.

In order to take advantage of the potential of teams, it is necessary to give them autonomy and responsibility, not to their individual elements but to the team. One of the most frequent mistakes is to give instructions, positive and negative remarks, recognition and rewards directly to members of a team. A team cannot be expected to work effectively as a team if there are direct professional communication between managers and specific team members. Communication should be made to the team as a whole or to its leader, who will then transmit it to the team. Everything that happens, good and bad, should be the responsibility of the team and not some of its members. It is up to the team to handle the responsibilities of each of its members inside the team boundaries. There should be no assessment and monitoring of the performance of each member of the team by the management. Management can only monitor the performance of the team as a whole. It is up to the team, if it so wishes, to carry out some type of performance evaluation of its members internally and if it wishes to recognize any of its members internally. The teams must be given autonomy to make decisions according to their state of maturity and according to what makes sense for the management, but the degree of autonomy must grow as the team becomes more mature. In order to foster team spirit, it is important to create moments of socialization outside the work environment so that team elements can get to know each other better and create bonds and become more committed to the team. Some degree of competition can be interesting to boost team spirit. Mature teams can make decisions about their elements' schedules and breaks, safeguarding the team's work commitments. A very big advantage for the manager of a certain production area when managing teams instead of managing people is that the number of teams is much smaller than the number of people. Measuring and monitoring the performance of teams is much easier than doing the same for each employee. The area manager is no longer concerned about whether each of his employees arrives on time or takes too long breaks, because this becomes the responsibility of the team. The team only has to do what is asked by managers and it is up to the team to manage its members. The existence of autonomous and self-managed teams brings enormous advantages for its members, while it brings enormous advantages for the management. It is definitely a win-win solution.

8.6 Team Leader

In all cases that we know of where there is teamwork formalized in organizations, there is a figure of a team leader. The role of the team leader is at least to lead the team meetings and to serve as a preferred means of communication between the representatives of the CI team and the area managers. The team leader is also expected to represent the team when necessary and to promote within the team the principles of excellence in organizations and the organization's values and vision. A good team leader can foster team spirit and the willingness of all its members to do better and be proud of the team they belong to. After dealing with teams for many years there is a somewhat harsh reality that seems to be evident, the performance and spirit of teams depend heavily on the leadership competence of their leader and teams do not always have a good leader. This means that in the teams you create in your organization you cannot expect them to find all a good leader. Some of the teams you create will work well and others will not work that well. Your role as CI promoter will be to be aware and identify which teams have problems and try to find ways to get around those problems. Sometimes it is enough to visit these teams more often, but there are cases where it is necessary to change their composition by changing elements between teams. The truth is that there will always be teams with lower performance and some with real problems.

The issue of assigning the role of team leader to a team member has some aspects to be taken into account. There is a question that often arises which is whether the team leader should be chosen by the team or appointed by the area manager? There are cases where the choice is quite natural, but there are cases where phenomenon such as "cockfighting" may come into action. The truth is that you need to be careful. Another question that arises sooner or later is whether the team leader, because he or she has a more relevant role than the others in the team, should or should not have a higher salary. This subject can generate some difficulties. If the organization decides not to pay a higher salary to the team leader then the motivation to be a team leader may not exist in some potential members. It is all very debatable and we do not even have a strong opinion about it. To get around this problem, some organizations adopt a slightly softer designation for this role of team leader so as not to place too much importance on this role. Some organizations give him or her the designation of "pivot", others "delegate" and in some cases even promote the rotation of this role in order to dilute responsibility and allow the development of skills in all team members. In any case, in this book no recipe is provided on how to create successful leaders and teams although there is plenty of literature on this subject for the reader to explore. Some knowledge has been created and disseminated in relation to football teams and other sports but we think that where one can probably learn more, due to the proximity of the context, is with the knowledge that has been created in software organizations. These organizations have learned to take advantage of teamwork in very effective ways. The use of Scrum and DevOps as team and project management methodologies has proven to be very effective. There is reference to these methodologies in chapter 7. We suggest that you make the best use of the knowledge that has been shared about these ways of organizing and managing teams.

8.7 Operational Team Meetings

The operational team meetings are events of great importance for the sustainability of continuous improvement. The frequency of these meetings is commonly proposed in this book's reference models (*Kaizen*, *Toyota Kata* and *Scrum*) as one meeting per day. In the *Kaizen* model, these meetings are even called *Daily Kaizen*. Typically, the meeting has a default agenda of 5 minutes that is followed point by point each day. In general terms, the objectives of these meetings are typically as follows: preserve and improve the cohesion and motivation of the team, become aware of the status and evolution of performance indicators, plan the day's work, monitor ongoing improvement actions and report occurrences if necessary. In some organizations, such as the case of *Company_C* presented in the chapter zero, which belongs to a large multinational in the automotive industry, the meeting agenda is specific for each day of the week. In the case of this organization, the schedule of team meetings is different each day of the week, although the following topics are common every day:

- A quick review of what happened the day before in terms of plan compliance and in terms of quality issues;

- Awareness of the production and training planned for the day;
- Warm-up exercises

And depending on the day of the week, there will also be space for other topics such as suggestions for improvement, 5S activities, remembering quality principles and other matters, which will be on the agenda depending on the day of the week.

For these team meetings to be effective, teams need to be trained, coached and monitored very closely in the first few weeks. After that, they must continue to be coached and monitored but the same dedication will not be necessary. It is necessary for someone with experience to be present in the first meetings, giving suggestions and comments on the content of the agenda and on the way the meeting was conducted. In the beginning, the agenda should have only two or three simple points to be accomplished at the meeting. Our suggestion is to schedule attendance, discuss the previous day's performance and make a brief plan for the day. As these points are covered in less than 5 minutes then other points can be added to the agenda such as identifying problems or simple improvement opportunities and placing them on the team board. In the beginning, these meetings tend to take more than 5 minutes and some issues remain unaddressed but as the team gains experience it becomes more effective in time management and ends up being able to do the entire meeting in about 5 minutes for the most part of cases.

It is important to repeat as often as necessary that the meeting cannot include any kind of blaming or any kind of judgment on team members. The spirit of the meeting should always be positive and motivating. The team members should always be encouraged to try to identify something that may be preventing them from doing their job better, with less effort and less risk. In fact, to make it clearer and more precise, look for sources of "*Muda*", "*Mura*" and "*Muri*", concepts already presented in the chapter zero of this book. Employees must be encouraged to seek to improve their work, even if it is just a little bit. The "a little bit" aspect is important to be underlined in order to give importance to improvements, however small they may seem. Employees should then seek to identify, in their routine activities, elements such as:

- Bending down or reaching for parts, equipment or tools;
- Small movements or transport that can be reduced or eliminated;
- Waiting for parts or equipment to complete cycles;
- Frequent occurrences of defects;

- Tasks that carry some risk;
- Very repetitive movements;
- Lifting excessively heavy loads or in difficult positions;
- Stress-causing situations;
- ...

It is desirable that all team members gradually become aware that in reality they can, and should, play a leading role in the way the work is done and how the area should be organized. Everyone should be learning that they can make small changes in their work area and in their workstations that result in improvements in their working conditions, however small they may be. The team must be trained to assume the principle that daily work includes the traditionally defined production or office tasks but also the responsibility to continuously improve the jobs and the entire working area of the team. This kind of awareness and learnings are only achieved with frequent visits by CI experts to talk to the teams and transmit these lessons to them.

The well-being of team members must always be valued by managers because respect for the human person has an unquestionable intrinsic value and also because people who feel good can produce better, with more quality and in a safer way. It is important to remember what was said in the first article published in English about the TPS (*Toyota Production System*). In this article it is said that one of the two basic concepts of TPS is "*treating workers as human beings and with consideration*" (Sugimori, Kusunoki, Cho, & Uchikawa, 1977). It is important that the entire organization breathes awareness of this concept and that it creates standards, routines and practices that materialize it and that it gradually transforms into the organization's culture.

At *Company_I*, previously presented in the first chapter, the operational teams hold a daily 5-minute meeting at the start of their shift. In this daily shift start meeting, a standard agenda is followed with the following points:

1. Registration of attendance;
2. Update of KPIs;
3. Presentation of the work plan for the current day;
4. Follow-up of ongoing actions;
5. If necessary: accident report; quality issues and delayed actions; information about sample production; new standards.

The same 5 minute duration for daily meetings was observed at *Company_H* where those meetings are scheduled at the start of each shift for each team but may not occur depending on the decision of the team leader. The team performance indicators are updated whether or not there is a team meeting and the daily 5-minute meeting agenda is as follows:

1. Discuss how the previous day went;
2. Brief description of the work plan for the current day;
3. If necessary, other issues related to accidents and/or quality problems are addressed.

At *Company_B*, operational teams range in size from 9 to 30 elements. In this organization, it is assumed that the size of the team is a function of the size of the production unit in question. If a line has 30 people then the team is made up of 30 people and the line leader is the team leader. Each team meets for 5 minutes a day and the theme is fundamentally safety. In case there was an incident or accident the day before, that incident or accident will be discussed to avoid other cases. Otherwise, there will be actions to raise awareness of the existing safety and dangers.

Lipor organization, in Ermesinde (Portugal) has natural teams in all sectors of the organization, whether in production or in indirect or support areas. Natural teams can and do carry out disruptive improvement projects too and are managed in the same way as temporary cross-team projects.

Although in a very different sector of activity that is the software industry sector, other approaches can be found. At *Company_P*, operational teams meet when they feel they should meet and can do it several times over the course of a working day without any time limit or agenda setting. It is clear that in these teams there is a large component of intellectual work, but there are some lessons that the traditional industry can learn from some practices in this sector of activity.

8.8 Team Boards

The existence of visual instruments for the organization and management of a team is essential. The team board is the main physical artefact associated with the team management and it can even be said that it is the "face" of the team. A team board says a lot about the team and can be a tool for team cohesion. The team board is like the instrument panel of a piece of equipment or a vehicle. Simply by looking at that instrument panel a lot can be learned about how this equipment is operating. An example of a relatively

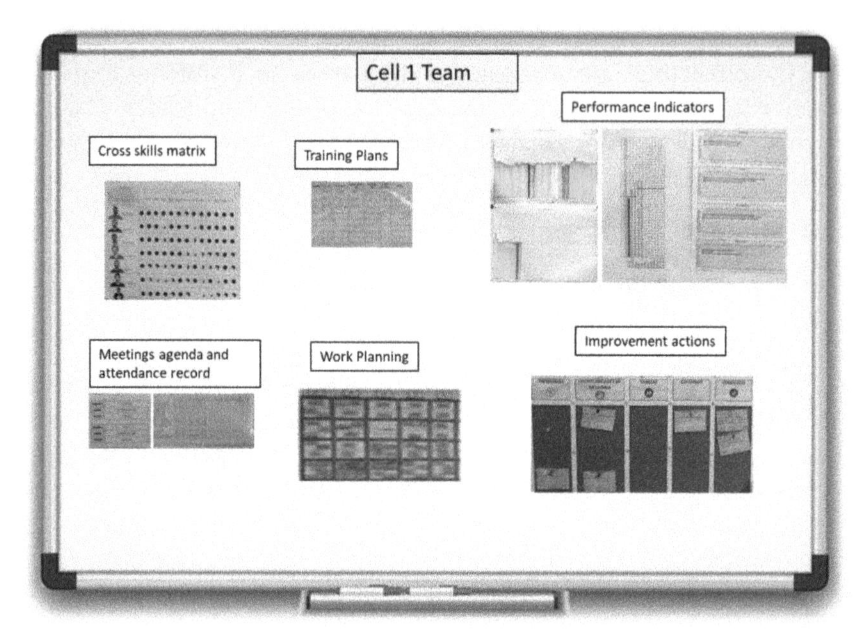

Figure 8.2 Team board example.

simple team board is shown in Figure 8.2. This team board presents some of the classic types of information found in organizations that take continuous improvement most seriously. We will present and clarify each of these types of information below.

The cross skills matrix is a representation (preferably with a graphical component) of the distribution of competences/skills of each member of the team. Typically a competency matrix includes the composition of the team, the competencies of each of its members (its versatility), the degree of mastery each member has in each of the competencies and which team members share each competency. It is important that there is not just one member with the capacity to carry out some of the team's tasks. It is desirable that there is always more than just one team member for each of the skills that are part of the team's responsibilities.

The training plan is another type of information that makes sense to be part of the team board because it demonstrates the team's awareness of its growth and development. This plan presents the training that is defined in a timeline for the team and for each of its members in order to respond to possible gaps in the team's versatility, growth and resilience.

An area with performance indicators and targets is probably the most common element in operational team board. There is a famous quote that says that you cannot manage what you cannot measure. Measuring is having a valuable answer about the effect of our actions and decisions. Teams do not always accept with pleasure and without reservations when it comes to discussing the subject of measuring their performance. It is often seen as a form of control and therefore the matter must be approached carefully. It must be clearly communicated that the idea is not for performance to be achieved at the expense of more effort or more stress, but rather the opposite. The goal is to find ways to do the work with less effort, less stress and in a more pleasant way. Measuring and monitoring performance is important for the team to see the impact of their improvement actions and also for managers to have no doubts that even working with less effort they can objectively achieve better performance.

> Six months after creating about two dozen teams in a very traditional small company with just over a hundred employees we conducted a survey and many of the responses we had from that survey were somehow interesting and somewhat unexpected. The answers that appeared referring to the daily monitoring of the indicators said things like: "I like the indicators because now we know how far we are going", "in the past we had no idea, but now we know how much we can produce per day". I remember that one of the teams even started to monitor some indicators 4 times a day so there would be no surprises at the end of the day.

Regarding the measurement frequency it is necessary to take some aspects in consideration. In almost all cases people tend to assume that measuring performance once a month or once a week is sufficient. In fact, the greater the frequency of monitoring, the more quickly deviations are detected and corrected. In most large and well organized multinational factories, the main production indicators are constantly monitored. There are panels in assembly cells and lines constantly showing the deviation between predicted (planned) and the actual performance. Of course, there are indicators that, as they do not vary constantly, it makes no sense to monitor continuously, such as the number of days without accidents or the number of customer complaints. In the first case just updating once a day and in the second case once a week should be acceptable for most organizations. The frequency of monitoring depends on how often the indicator varies and the impact of deviations from that indicator on the organization's overall performance. It is a good practice to have defined objectives for the indicators that each organization and each production unit defines to monitor and it is also good practice to indicate in

Table 8.1 Example of a work plan for a team.

	Monday	Tuesday	Wednesday	Thursday	Friday
Manuel	Meeting	Task 1	Attendance	Task 1	Task 2
	Attendance	Training	Task 2	Attendance	Attendance
Maria	Meeting	Task 5	Holiday	Holiday	Holiday
	Task 1	Training	Holiday	Holiday	Holiday
Ana	Meeting	Attendance	Task 1	Attendance	Task 2
	Task 1	Task 3	Training	Training	Training
Miguel	Meeting	Task 3	Task 1	Task 1	Attendance
	Training	Attendance	Attendance	Task 3	Task 4

green when the indicator is better than objective and in red when the opposite happens. This seems to be the norm in the overwhelming majority of organizations. It is marked in green when the result is better than the objective and in red when the result is worse than the objective (see the indicator area shown in the upper right part of the team board shown in Figure 8.2).

The standard meeting agenda and attendance registration is also quite frequent in continuous improvement team boards. The team meets in front of the board and having the agenda in front of everyone helps to manage the meeting. The registration of attendance and any delays is just to understand the history and give a formal emphasis to the meeting.

The format of work plans varies a lot from team to team and in some cases, especially in production, this plan exists in another system outside the team board. In indirect areas, it is common to adopt a schedule for each day of the week. In this plan (see Table 8.1), for each team member, the way in which their tasks and commitments will be distributed each morning and each afternoon of each day of the week is defined. This plan is defined for example in the meeting that takes place every Monday morning. Based on this plan, it is easy to have an idea of what will be produced during the week and at all times all the elements know what they are supposed to do and what each of their colleagues is doing.

In the direct production areas, the work plans for the operational teams are, in most cases, an ordered list of production orders for batches or parts to be produced on the day or shift in question. This list must be followed by the order of priority given by the local supervisor or manage and there are, unfortunately, frequent cases in which this ordered list is changed several times throughout the day. There are also cases where the decision on which order to process is up to the team itself in order to optimize, for example, tool changes on the machines. Ideally the decision on which order to process next

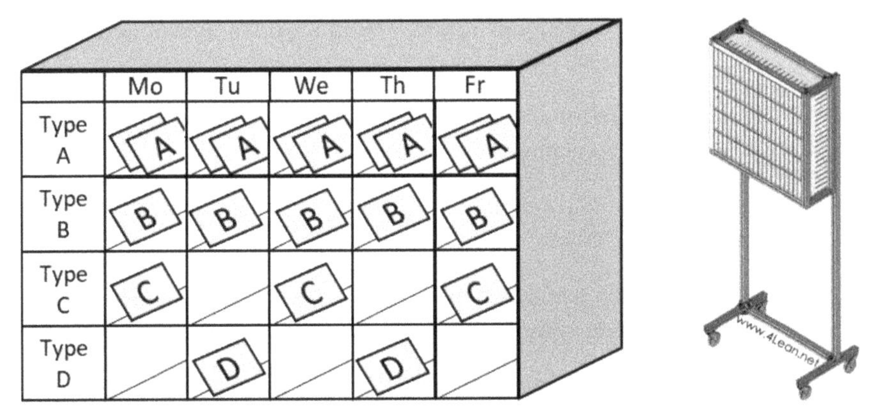

Figure 8.3 Examples of heijunka boxes (courtesy of 4Lean.net).

should have to do with maximizing the flow, i.e. for example the order that is needed earlier in the next process.

In some cases the team uses another artefact outside the team framework where production plans for the team are defined by the production management system. In organizations that adopt production flow creation principles, these plans are materialized in the well-known Heijunka boxes with typologies equivalent to the examples presented in Figure 8.3.

The space reserved on the board for improvement actions serves to keep the status of the improvement actions created by the team within the control of the team. The most relevant information to be clearly identified in this space is the description of the action and the state it is in. It is common to use the phases of PDCA cycles to identify the state of action, but there are other ways. In the table shown in Figure 8.2, the improvement actions are described in a small form that changes columns as its status progresses. Possible states can be for example: "Proposal", "In Evaluation", "In Implementation" and "Completed". In some cases, in addition to proposals for improvement, identified problems can also be put forward without a proposal for a possible improvement action.

8.9 Project Teams

As has been said frequently throughout this book, the effort for continuous improvement must be widespread throughout the organization and must be part of everyone's daily routine. Despite this, there may be a need for disruptive improvements that go beyond the autonomy of existing teams. There may be a need for structural and/or technological changes that cannot be carried out by existing continuous improvement routines. Possible examples are:

- Changes in production areas affecting more than one operational team to better respond to changes in the demand pattern,
- Increased demand and consequent need to increase the production area and transform the layout,
- Introduction of new equipment that transform production and teams with a view to improving some indicator or set of performance indicators,
- Improvements with rearrangements in the departmental structure,
- Improvements with rearrangements in common areas,
- Improvements with changes in communication systems,
- Improvements with disruptive changes that affect the entire organization globally,
- Etc.

In these cases there may be a need to join a temporary team, people from different departments with specific skills that best serve the needs of the project in question. These teams whose intent is to achieve disruptive improvements are called project teams and only exist for the duration of the project for which they were created. These improvements usually result in new work standards of behaviour that are normally expected to significantly gain in performance, competitiveness or in the well-being and satisfaction of employees.

At the Bosch factory in Braga, these teams are endowed with great autonomy, are governed by the principles of PDCA cycles and are constantly advised by the *Value Stream Mapping* manager, who acts as a coach throughout the process. Every three months, three production areas are chosen as the focus of attention. This choice is carried out by management personnel in a joint meeting where all possible choices are displayed on a magnetic whiteboard, together with the possible gains from their implementation. It should be noted that, in this whiteboard, in addition to the possible gains, the suggestions given by employees for that respective area are also displayed. For each project is typically defined a VSM of the current state and a VSD of the future state (*Value Stream Design*) that is desired. Once the improvement has been made, new working standards are developed and then implemented and validated in the field. This is followed by a validation of the benefits achieved with the new way of working and new standards.

In the case of another organization, *Company_B*, there are three different types of disruptive improvements: there are so-called improvement events, improvement projects, and quick actions. Improvement events are established in advance by the management team, and are characterized by a duration of one week. During this week, three multidisciplinary teams, specially formed for this project, are working on a problem whose solution is unknown. Each team is assigned a different objective factor:

$$Objective = People \times Costs \times Quality \times Throughput$$

The idea is that any improvement will have to be translated into improvements in people, a reduction in costs, an improvement in quality, an increase in the quantity produced, or a combination of these factors. In that week of the event, the team members are dedicated exclusively to the problem at hand. These improvement events are equivalent to what are commonly known in many organizations as *Kaizen Events*.

The second type of disruptive improvements arises when the problems to be solved or the improvement opportunities to be studied require, in the first place, more resources and more time than the improvement events presented before. These types of projects are more similar to the ones mentioned above referring to the Bosch Company and which are more commonly referred to as disruptive improvement projects. The third type of improvement that *Company_B* offers comprises easy-to-resolve obstacles, they are called "Quick Actions" and only 1 or 2 people are allocated to these such "mini interventions" for improvement.

8.10 CI *Instructors* in Operational Teams

To get the operational teams to put their full potential at the service of CI on the ground, it is necessary that some people, with specific skills, accompany them on a regular basis. These people must on the one hand have skills in implementing continuous improvement and implementing the principles associated with models of excellence in organizations such as the *Lean* philosophy or *Shingo* model and on the other hand have some skills in communication, teamwork and creating empathy with the others. There are several possible names for the role of these people in training and monitoring the operational teams in the CI effort. It is not easy to decide which designation is best, and that could probably be due to the fact that we do not have one that completely satisfies the idea that it is desirable to be conveyed. The reader can choose one of the following: "Coach", "Animator", "Instructor", or another that judges better than these. The role of this CI "animator" is to develop the best possible skills and attitudes in the team, such as teamwork, team spirit, continuous improvement skills, skills in materializing some principles of the *Shingo* model, *Lean* philosophy, or the *Toyota Way*, willingness to improve performance, critical thinking, willingness to grow as individuals, and other skills and attitudes in this line of thought.

These CI animators play a very important role in CI's sustainability because in fact they are the real responsible for its success. The person or people who get this role, in our opinion, should have other tasks and responsibilities than just the role of accompanying the teams in CI. It is better to have several people playing the part-time CI animators role than just one person dedicated exclusively to that role. While this is our case, it is often the case that early in the process of developing a CI system these responsibilities are assigned to a single, full-time person.

8.11 Continuous Improvement Command Centre

The existence of a room for CI management activities is a very common practice in organizations with CI systems. There can be various purposes and designations for this type of room. This type of room is often referred to as the "*Obeya*" room, a Japanese word meaning "Large room". Another popular designation is called "war room" which comes from practices used by military personnel. The designation "*war room*" although its meaning is understood as being the command center, is probably a designation too strong to be used in some cultures. Probably the best designation would be the continuous improvement command center. This CI command room or center has the following purposes:

- To keep visual information and charts/monitors with global CI indicators updated;
- To be a place for continuous improvement team meetings;
- To be a place for meetings of disruptive improvement project teams;
- To maintain boards with visual information and indicators of ongoing improvement projects;
- To be a place that can be used for eventual meetings of operational teams to resolve some of their specific problems;
- To maintain a structured problem-solving framework for any team to use as a guide in solving their problems;

- To be a place for short CI trainings;
- Other CI related purposes.

Regarding the framework for structured problem solving, it is common to adopt a specific board with the pre-marking of the fishbone of the *Hishikawa* diagram or any other pre-marking for problem solving such as the format of the A3 reports. This type of support can be important in helping the team create the right environment for systematic discussion and problem solving. Other organizations use other strategies such as *Company_P*, dedicated to software production, where the walls are all painted with a specific ink to be able to write and erase as if it were a normal whiteboard. Teams can meet near any wall whenever they feel the need to draw diagrams, or any other method to support the discussion.

8.12 Suggestion Systems Integrated in CI Routines

The use of the potential creativity that exists in all employees in an organization is something that has been neglected for decades in our companies and organizations, with huge losses for our economy and for the population's standard of living. That way of thinking in which some people are paid to think and others are paid to perform is believed to be completely out of date and very ineffective. It is everyone's interest that everyone participate with their creativity for the individual and common good. Not using this creative potential was even considered to be the eighth waste, in the famous book "*Lean Thinking: Banish Waste and Create Wealth in Your Corporation*" by James Womack and Daniel Jones (Womack & Jones, 1996). In addition to the classic 7 wastes that were defined in the Toyota Production System and referred to in the first chapter of this book, this eighth waste was described by the authors as the non-use of creativity by all employees. This waste can be described as being all the potential that the organization loses in time, ideas, skills, improvements and learning opportunities by not engaging its employees or simply listening to them.

An important leader from the *Company_H* plant in Penafiel (Portugal) started the journey of continuous improvement almost 20 years ago and once described to me his journey in continuous improvement as follows: "In the beginning I spent a lot of time identifying problems and improvement opportunities, I was the only one thinking and the evolution was of little relevance. Gradually I passed on the task to all the employees, and they started to identify the problems and the opportunities for improvement. I gradually stopped being the only one thinking, and the number of improvements identified and implemented, as well as their positive impact on the overall performance became much, much more relevant. This gave me more time to take that all-important 'helicopter view', to think big and to better plan the future of the organization and the people who work there".

In our opinion, suggestions management should not happen outside the CI system adopted by the organization. Suggestion management should happen in an integrated way with the continuous improvement system. It makes sense that the suggestion systems are embedded in the CI system routines because the suggestions for improvement are obviously closely linked to the concept of continuous improvement. The identification of problems to be solved and the identification of improvement opportunities should be a task of the operational teams, as it was already partially mentioned in this chapter. The search to improve the team's performance is part of the daily work and as such the need to suggest and implement improvements should be part of the teams' life. The suggestions related to the team's work area should be published on the team's board and followed up by the team itself and by the team responsible to manage the CI system in the organization. It should be that CI management team (if that is the mechanism) responsible to validate and implement the improvements that are outside the sphere of competence of the operational team itself.

Some suggestions can be implemented by the operational team itself but others require the skills of others. Changing the sequence of operations or the operating mode to carry out a task can in many cases be carried out by the operational team itself, but other improvements, such as changing the programming of a machine or installing an automation, may require the intervention of people outside the team.

Another aspect worth mentioning is that printed forms are usually created for employees to describe the problem or opportunity for improvement they suggest. In our opinion you should start with something simple and small. The most important is the description, date and author. In *MoldartPovoa* Company, for example, there is a printed form in one color to identify problems and an equivalent one in another color to identify suggestions for improvement. Suggestions or problems should be placed on the team board and the continuous improvement team should assess, in the field, what action should be given to each suggestion. Even if the description is not easy to understand, being in the team's production area makes it easy to contact the author of the suggestion so that he or she can better clarify exactly what he or she wanted to say. These moments of communication between representatives of the CI team and employees should be used to give positive reinforcement to the team and motivate them to continue contributing with excellent ideas. It is common that, in these moments of conversation with employees and giving them genuine attention, it is when they normally provide super interesting ideas and insights that should be implemented even in other areas of the organization. In this case the temptation is to tell them to write it down and put it on the team board to be treated according to the prescribed procedure ... but there is a better way. In these cases, it is very effective to write ourselves (CI team members) what the contributor said. Employees sometimes feel unsure about the value of their ideas, whether or not it's worth the trouble to write an improvement idea. They are often not sure if it will be good enough to be expressed. For these reasons it's good to be someone from the CI team to put that suggestion down on paper. With this approach employees become more confident because in that way, at least someone already thought the idea was valid since a member of the CI team, normally a manager, has volunteered to put it on paper.

Although the identification of problems and the generation of suggestions for improvement in the work areas associated with each team can be generated within the team itself, there are common areas, procedures and rules that are outside of team control and responsibility. In addition, there are also interface areas between teams. Some teams provide materials or services to other teams and therefore not all the room for improvement is naturally covered by a particular team. For this reason, many organizations have a suggestion system parallel to the suggestion system within each team. Anyone who wants to make suggestions about border areas, common areas (dining rooms, bathrooms, corridors, outdoor spaces, etc.) and even how other teams should work can use this parallel suggestion system. These suggestions must be evaluated by a panel of managers and the verdict must be given in good time to the author of the suggestion. The Bosch factory in Braga is an interesting example where there are these two spaces for suggestions, a system for local suggestions in the team and another system for general suggestions. For this type of suggestions there is a computer system so that employees can register their suggestions and earn credits with them. This issue of earning credits is discussed next.

8.13 Recognition / Award

Whenever a suggestion system is initiated, it is important that suggestions are responded to in a timely and appropriate manner. Everyone who puts forward a suggestion on their own initiative likes to be recognized for that effort and exposure. It is important to show recognition and give positive reinforcement every time someone makes a proposal. Whatever it may be. Even if the suggestion is not aligned with the strategy and culture of the organization it is necessary to explain it in an adequate and friendly way to the employee. What we want is that this employee continues motivated to create other suggestions that next time will be aligned with the strategy, values and culture of the organization.

There are sentences that are always well received by people and teams who make suggestions and implement improvements. Of course, each manager and each leader has their own way of expressing themselves, but just to give you some examples, here are two examples of such sentences:

"Congratulations on the excellent improvements you have been making"

"This idea of yours is magnificent".

If this initial phase of responding and giving positive reinforcement to the teams or people making suggestions fails, the whole suggestion system is at risk of resulting in a huge failure. Preparing a robust system that can deal with suggestions is vital for the survival of the suggestion system itself and for continuous improvement as a whole. Managers who make the effort to implement a suggestion system should be prepared for the time they will have to spend evaluating all the suggestions that will be generated. In the beginning everyone is available to evaluate the suggestions that are created but quickly no one has the time. Remember that at the beginning of any project everyone is motivated and everyone contributes but the real difficulty is its sustainability in the medium and long term. When you think about implementing a suggestion system you have to think very well how to keep it alive and that always requires hours of work from managers. Be prepared for that. The good news is that the time spent implementing improvements has a much higher return than the time you need to spend every day to carry out the tasks that exist because the improvements were not implemented. This idea is present in the famous metaphor that time sharpening the axe is saved later cutting down the trees, presented in chapter 6.

But back to the core issue of recognition and reward, should incentives or rewards (monetary or otherwise) be given to employees for each improvement suggestion made? This is a question that is often asked with regard to suggestion systems. Some aspects of this question were already addressed in chapter 1, especially about the practices of Toyota and some American organizations. While in Toyota there was a fixed value around 10 USD per suggestion in American organizations it was typical to give

10% of the achieved gain generated by the implemented suggestion. The Bosch factory in Braga used to give until recently 5 credits for each suggestion, which could be used to buy Bosch products in the company shop, whatever the suggestion was. This led some collaborators to change their focus to earn points instead of the pertinence of the suggestions. The system was abandoned and all suggestions must now go through the local supervisor's approval and only he can submit them in the suggestion system. Furthermore, a bonus well above the 5 credits is foreseen, but only for suggestions that are implemented.

It may be relevant to present here how MoldartPovoa deals with this issue of acknowledging improvement suggestions and implementations. This organization, after some attempts, decided to give a prize for each improvement implemented. Notice that the award is not given for each improvement suggestion but for the suggestions that are transformed in implementation. The prize is given at the end of each year, it is 50 euros for each improvement implemented and it is given to the team instead of to the author of the suggestion. The idea of this system is to help reinforce team spirit.

There is a very interesting and famous book entitled "*Drive: The Surprising Truth About What Motivates Us*" by Daniel Pink (Pink, 2009) that addresses, with substance, this issue related to the role of incentives in creativity and work engagement. Daniel Pink, based on the results of many published scientific experiments suggests, in summary form, that incentives can be counterproductive when creativity is desired. There is a very interesting experiment known as "*Duncker's Candle Problem*" (Duncker, 1945) which shows us a very interesting human aspect. In the initial formulation the subject is placed before materials placed on a table as shown in para a) of Figure 8.4. It is about a candle, a matchbook and a box with tacks. The challenge is to find a solution so that the candle is somehow connected to the wall and the wax does not fall on the table top.

After a few tries for 5 to 10 minutes most people can find the solution shown in part b) of the same figure. The key to solving this problem is to

Figure 8.4 Duncker's Candle Problem. Adapted from http://whatismotivation.weebly.com/the-candle-problem.html.

overcome what is called functional fixation. People assume that the function of the box is to contain the tacks and do not quickly see another function for the box. If the problem starts with the configuration presented in part c) of the figure, people solve the problem much more quickly because the problem of functional fixation is not so strong anymore.

Another experiment was conducted but this time with financial incentives for those who solved the problem in less time. How much less time does the reader think that people who were given financial incentives took longer than people who were not given any financial incentives? The truth is that according to what is published in Daniel Pink's book (Pink, 2009) these people took an average of 3.5 minutes longer. The incentive did not increase the intellectual and creative nature, on the contrary, it came to cloud and limit it.

Apparently creativity seems to be more natural when it is taken as a personal challenge without the external "pressure" generated by an incentive. Since much of what is intended with the suggestions is based on people's creativity, it is a little strange to add something (an incentive) that seems to limit creativity. Also according to the aforementioned author, salary incentives are more effective when it is intended to increase productivity in less creative work, namely repetitive operations without the need for intellectual disruption.

We believe that the study reported above has its specific characteristics and context and we cannot assume that in a organization, incentives have this perverse effect. In any case, some care has to be taken in assigning incentives to creativity because there is some danger that it will be counterproductive. Regardless of this, we believe without any doubt that the organization's recognition of the teams' positive results must be expressed in a clear, transparent and in a publicized way. There may or may not be a place for rewards, but there are some studies and experiences that argue that these rewards should be symbolic. The use of monetary rewards associated with the gain for the organization, generated by the ideas, can be complex and generate some discomfort among colleagues and some feelings of injustice. Furthermore, it is important to say that we believe that the recognition and/or reward effect is more positive if it is attributed to the team than to individuals as it can also serve to help reinforce cohesion and team spirit. There are organizations that create the figure of the team of the month and others the employee of the month as a way of acknowledging the organization for contributions to the good of the organization or for the relevant role they played in the organization. Our opinion, because we defend teamwork and the synergy that can be created with the appropriate team spirit, recognition should be given to teams rather than individual people.

8.14 Continuous Improvement System

The CI support and management system is an essential element for the proper functioning, synchronization and alignment of Continuous Improvement activities with the objectives established by the strategic vision. This element is called "System" because, in addition to the CI team, it also includes all the routines, necessary artefacts and standards to be followed for the CI to be stable and sustainable. The purpose of the CI system is to train and assist teams in the field, ensuring that the effort for continuous improvement is constant, integrated into the organization's operations and carried out by all.

For you who is starting or revitalizing continuous improvement journey in your organization, it is advisable to create a continuous improvement management and support system led by the continuous improvement team we talked about at the beginning of this chapter. Members of that team may or may not devote all of their time to CI tasks. In fact, it is even advised that no one in the organization devotes all their time to CI. CI's responsibility should only complement the exercise of other positions in the organization: we can, for example, have someone who works in the Human Resources of an organization for x% of their time, spending the rest of their time on training and assisting the teams in the scope of Continuous Improvement.

As an example, in the public organization that was previously referred to as Lipor, the CI management and animation team is a multidisciplinary team consisting of 7 people (1 person for each different department) known as "*Kaizen Team*". This designation is the result of the *Kaizen* Institute's involvement in the initial development of the organization's continuous improvement system. The 7 people who are part of this team dedicate around 40% of their time to continuous improvement, using the other part of their time in the position assigned to them in the respective department. The motto of this team is very clear! "*To be a driving force for continuous improvement at Lipor*", and as one of its members says, "the idea is that the *Kaizen* team ensures that the philosophy of continuous improvement is omnipresent throughout the organization".

In another example of CI system, in *Company_B* (part of a large multinational) the team responsible for leading the CI, under the name of "*Steering Committee*" is made up of top management people (department directors), and is responsible for outlining the continuous improvement events for a one-year time frame. From a set of possible disruptive improvement events, which were obtained by suggestions from employees and managers, those that apparently benefit the organization the most in the following four distinct indicators are chosen:

- Working conditions, safety and employee motivation;
- Lead times;
- Number of defects;
- Operating costs.

Interestingly, this organization places greater emphasis on the advances that can be achieved through disruptive improvements associated with project teams than on the advances that can be achieved through the continuous improvement achieved by the operational teams. For the first case, there is a disruptive improvement event planning and for the second case this organization includes two mechanisms. The first mechanism is based on daily visits to the production area (*Gemba Walks*) by the CI team for this purpose. This team, called the *Steering Committee* as previously referred, involves 12 people and takes a "walk" throughout the factory every day. As the *Gemba Walk* take place during working hours, the line workers are given headphones so that they can hear everything that is said during the 10 minutes. In the particular case that a major problem was detected the day before, the person responsible for that problem area is called to the meeting and a brief analysis of what happened with him is made. The second mechanism focuses on the work assigned to two members of the *Steering Committee*, which consists of providing CI support to the operational teams in the field. These two people specialized in CI, excellence in organizations and with in-depth knowledge of the organization's CI system and its strategic objectives, are dedicated exclusively to continuous improvement. They use 100% of their working time to support teams (operational and project) and to analyze the production process, supporting and substantiating the improvement activities. This is a case of an organization where there does not seem to be any formalized routine for the contribution of employees to the continuous improvement effort.

There are still other examples of "hybrid" systems or models that assign control and support of CI to more than one entity. An example is the case of *Company_C* (another company from a large multinational group) which combine, in addition to a CI team, managers who individually contribute in the effort to support the teams in the field. The CI team is responsible for monitoring and ensuring the maintenance of progress made following improvement events. This team meets monthly with the purpose of taking pending decisions and analyzing ongoing projects. The department directors and factory manager are responsible for supervising the processes in the field, taking on the role of *coaching*, in order to obtain the contribution of the operational teams with small suggestions for improvement. These managers who act as *coaches* interconnect and assimilate CI functions as part of their responsibilities and daily routines, always seeking to awaken the motivation of employees and, whenever applicable, recognize their good performance. In addition, they are also responsible for taking decisions on the suggestions made by employees and managers, which may be the following:

- They can choose to proceed with its implementation, being a simple action;
- Reject, duly justifying;
- Propose the execution of an event for disruptive improvement.

A CI system for a SME should include, for example, entities, artefacts, routines and standards of the type shown in Table 8.2.

Table 8.2 is intended to condense the main entities, artefacts, routines and standards that we suggest exist in an SME CI system. The solution to be found in each organization may be a little different as there will not always have to be project teams but the other entities are probably the least that can guarantee some success in the CI journey. The artefacts presented are only proposals but we suggest that at least these exist. When we refer to "meeting space" it does not mean that there should be a room reserved for this exclusive purpose, but it would be good to have a room dedicated to the CI system. This room, which appears in the table with the name "CI Command Centre" (or *Obeya* Room) is where the main team boards of the CI management team, the team boards of the project teams and equipment that can be shared by everyone in their CI tasks. This room would also serve for operational teams to meet to solve problems or to hold meetings with very specific purposes that made sense to be outside their work area.

References

Bastos, A., & Sharman, C. (2018). *Strat to Action - O Método KAIZEN™ de levar a Estratégia à Prática*. Kaizen Institute.

Duncker, K. (1945). On problem-solving. *Psychological Monographs, 270.*

Monden, Y. (1998). *Toyota Production System: An Integrated Approach to Just-in-time*. Engineering & Management Press.

Ohno, T. (1988). *Toyota production system: beyond large-scale production.* (C. Press, Ed.) (3ª Edição). New York: Productivity, Inc.

Table 8.2 A proposal for an initial CI system.

Entities	Artefacts	Routines	Standards
IC management Team	• Space for meetings with a team board. • Monitor CI indicators.	• Weekly meetings. • Monthly meetings with the administration. • Daily or weekly visits to operational teams.	• Meeting standards. • Standard for visits to operational teams. • Problem solving standards. • Standards for selection of improvement suggestions.
CI *Coaches* and animators	• Table with the evolution of the assessment of the teams. • Registration of visits.	• Daily (or weekly) visits to operational teams.	• List of questions. • Team assessment standard.
Operational Teams	• Team boards. • Monitor team indicators. • Printed forms to make suggestions.	• Daily meetings. • Updating indicators. • Keeping 5S.	• Meeting standards. • Standards for 5S implementation.
Project Teams	• Space for meetings and team boards. • Project indicators monitor	• Weekly meetings. • Monthly meetings with the CI management team.	• Meeting standards. • Project management standard.
Several entities	• CI Command Centre (*Obeya Room*) • Framework for structured problem solving		• Problem solving standard. • Standard for use of space. • Standard for using the structured problem solving boards

Pink, D. (2009). *Drive: The Surprising Truth About What Motivates Us*. New York: Riverhead Books.

Sugimori, Y., Kusunoki, K., Cho, F., & Uchikawa, S. (1977). Toyota production system and Kanban system Materialization of just-in-time and respect-for-human system. *International Journal of Production Research*, *15*(6), 553–564. https://doi.org/10.1080/00207547708943149

Womack, J., & Jones, D. (1996). *Lean thinking: Banish Waste and Create Wealth in Your Corporation*. New York: Fee Press.

Womack, J., Jones, D., & Roos, D. (1990). *The machine that changed the world*. New York: Free Press.

Yasuda, Y. (1991). *40 Years, 20 Million Ideas: The Toyota Suggestion System*. Cambridge, MA, USA: Productivity Press.

Annex

Tools Frequently Used on the Path to Excellence

In this annex, tools that appear frequently linked to the Toyota-inspired excellence models are briefly presented, which have only been referred to throughout the book. In general, these are popularly called "Lean tools" but are also often referred to as quality tools and in some cases may also be referred to in a maintenance context. Much material is available about these tools on websites and YouTube videos, although not always correctly, reliably or accurately. Besides those sources, there are also many books on the market that present these tools in detail and in a correct way. For this reason, the aim of this annex is not to describe each of the tools in depth but only to give a brief idea of their purpose, their application and some suggestions on how to use them. Furthermore, indications are also given, where possible, on how you can deepen your knowledge on each of these tools.

Five Whys

The so-called "5 Whys" technique is very simple and is used by operational teams to solve problems, especially small ones, which occur on the ground. The idea is that because it is very simple, operational staff who do not have the time, tools and training for structured problem solving, can use it. This technique is based on asking "why?" five consecutive times, which in most cases is enough to identify the cause of the problem. For example, consider a spring nailing machine whose production rate was well below what it was supposed to be. The technique was applied as follows:

- Question: The machine producing so little, why?
- Answer: Because the machine is always jamming.

- Question: Why is the machine always jamming?
- Answer: Because the machine sometimes does not detect the existence of the spring.
- Question: Why the machine does not detect the presence of the spring?
- Answer: Because the sensor moves with the great vibration of the machine.
- Question: Why is the machine vibrating so much now?
- Answer: Because it has been moved and not properly positioned.

As this is a very simple technique, it tends to be overlooked, but in many cases, if used with the "right people" and the proper training, many problems can be solved. In fact, a very effective way is to hold a meeting with those people who know best the process where the problem is occurring (*e.g. equipment operators, supervisors, maintenance personnel and even managers*) and apply the technique. If you try to use this technique, you may find that the questions asked naturally depend on the knowledge of the problem you are analyzing. Like all techniques, a period of learning and improvement is required to achieve effectiveness in a large number of cases. This relatively structured way of problem solving has a more important effect than the technique itself. The fact of tackling problems in a systematic and formal way (*organizing a meeting of people following a defined technique*) is already an important step towards solving those problems.

Ishikawa Diagrams

Ishikawa diagrams (Ishikawa, 1985), also known as cause-effect diagrams, or, because of their visual similarity, fishbone diagrams, have been used for a long time to help break down a problem into an organized structure of possible causes. The engineer and university professor, Kaoru Ishikawa, worked in quality control in some Japanese organizations and proposed this type of diagram in 1943. These diagrams help to identify very comprehensively all potential causes and even include the possible impact of each cause on the problem (Figure Annex.1). Although they were created to solve quality problems, these diagrams can be used in many other areas.

Many problems do not have just one cause, but there may be several causes with different weights in the responsibility for the problem. When a problem is complex, the five Whys technique is not effective enough to solve it, but this type of diagram allows the team trying to solve the problem to

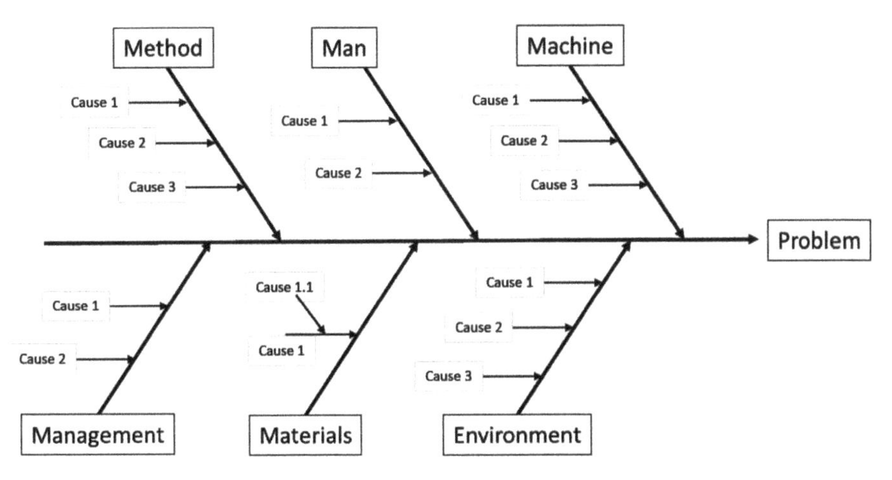

Figure Annex.1 Typology of an Ishikawa diagram.

do so in a systematic and comprehensive way. First, you start by identifying classes of causes and then identify possible causes in each class. The visual representation helps the team to build a detailed understanding of what might be the root cause or root causes of the problem.

A3 Reports

The A3 Report is a predefined report structure that occupies one A3 sheet, hence its name. This form, which incorporates the methodology, can serve two purpose: to serve as a record of an improvement achieved or a change made, or to serve as a guide and support in problem solving. There are some books dedicated to this methodology among which we can probably highlight "*The A3 Workbook: Unlock Your Problem-Solving Mind*" by Daniel Matthews (Matthews, 2010). In its structured problem-solving function, an example is presented in Figure Annex.2. In this real-life example of a organization, a problem was identified of an increase in the quantity of items in the manufacturing pipeline (WIP) and the duplication of production records. The amount of items on hold was generating many space management difficulties, difficulties in decision making regarding priorities and long throughput times. The solution found was to include a pull production algorithm that limited the amount of items on hold in each process and suggested at each moment what the best solution to take should be. The decision maker, based on the information of this suggestion given by the system, remains free to follow it or not according to his own interpretation of the best decision to take. An interesting game on this structured problem solving technique is

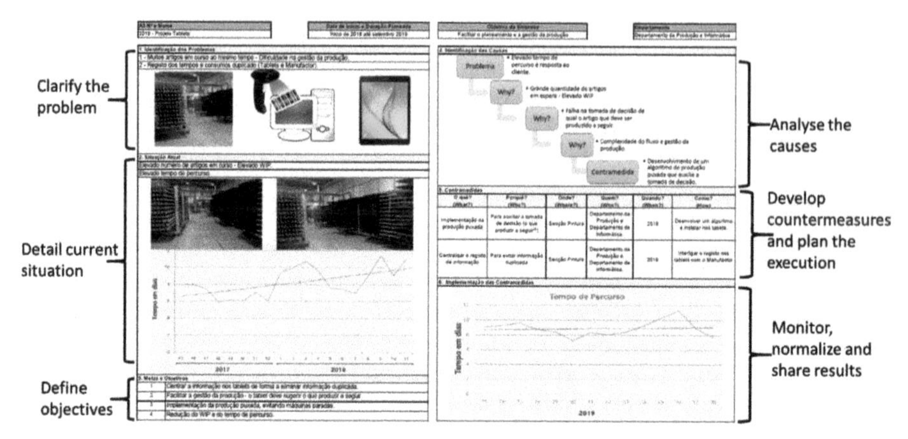

Figure Annex.2 Example of an A3 report in structured problem solving (courtesy of John Bicheno).

presented by John Bicheno in a book on Lean games entitled "*The Lean Games and Simulations Book*" (Bicheno, 2015). In this game book, the problem in focus is the case of a coffee shop that started to receive more and more complaints and to lose customers. John Bicheno is the author of several books on Lean and was for a long time connected to LERC (*Lean Enterprise Research Centre*). This structured problem solving technique guides the team systematically through 6 phases until the identification of possible causes and consequent countermeasures. In the first phase, the team starts by clarifying the problem in the best possible way, always using diagrams and as little text as possible. Then, in a second phase, the problem must be broken down into its components. This second phase can also be described as the detailed specification of the current situation. It is of utmost relevance to incorporate objective data from the components of the problem. Often different team members perceive differently the core of the problem and the constraints of its real context. It is very important to go to the field, talk to those involved and collect the most relevant details to dissect the problem. The better we know the current situation and the details of the problem, the easier it will be to detect its causes. In a third phase the team should establish the targets they want to achieve in terms of performance and in which period. Then, in the fourth phase, the team sets off to identify the causes. It is very common that when solving problems teams move directly to this phase without worrying too much about the previous phases, but this is a mistake. It is very difficult to discover the causes of a problem when it is not well known nor are the details of its context. Various types of methodologies can be used in this phase, one of them being Ishikawa diagrams, although in some cases the 5 whys technique works efficiently. Depending on the complexity of the problem and its context, different tools can be used.

The next phase, once the causes have been identified, is dedicated to the development of the necessary countermeasures to eliminate or circumvent the causes of the problem. It is also necessary at this stage to establish a schedule for the implementation of these countermeasures. Finally, you move on to the phase of monitoring the evolution, behavior and effect of the implemented countermeasures, stabilizing the processes and sharing the results so that other departments and sections can also learn from each project of this type. This technique is presented in more detail in chapter 6.

PDCA Cycles

The origin of the PDCA (*Plan, Do, Check, Act*) method (Rundle, 2019) is attributed to William Edwards Deming, who presented it in Japan in the early 1950s. Mention is also often made of Walter Shewhart who in 1939 had already formalized the cycles: Specification - Production - Inspection and was the source of Deming's inspiration. It would be acceptable to say that Shewhart's cycles and Deming's cycles are the formalization of the scientific method that had already been used for centuries, at least since Galileo. The scientific method is essentially a cyclical process in which the scientist plans the experiment, then executes it and then checks the result. Depending on the result obtained, there will be learning, reflection and a planning of a new experiment. We can also argue that this logic, although not formal, was present in human intervention throughout its technological evolution. This kind of approach must have been present since the development of the first chipped stone tools. Even believing that man's way of acting in many situations already informally incorporated some cyclical steps, one cannot take the credit from those who became aware of their existence and gave shape to them.

The PDCA cycles are very much associated with continuous improvement, as mentioned throughout the book. The way it is materialized is, for example, as follows:

- (P) Plan a new way of doing the operation;
- (D) Do the operation according to that new way;
- (C) Evaluate the result;
- (A) Make decisions: if the result is still not what you want, go back to (P).

The steps just mentioned will be carried out cyclically until the result is the desired one. PDCA cycles are used not only to achieve improvements but also to solve problems. Figuring out how to solve a problem can follow the steps indicated in the following example:

- (P) Plan a new approach, or a slightly different way, to solve the problem;
- (D) Execute according to that new approach;
- (C) Evaluate the result;
- (A) Make decisions: if the result is still not what we want, go back to (P).

We can have long-term and short-term PDCA cycles. An example of long-term PDCA cycles can be a marketing campaign. A new marketing campaign may have results only several months later, and this will trigger changes or improvements in the next campaign. On the short-term side is the example of increasing or reducing the feed rate of the cutting tool in view of the surface quality of a machined part. Furthermore, it is also possible to have short-term PDCA cycles within long-term PDCA cycles.

Hoshin Kanri

The translation of the Japanese expression "Hoshin Kanri" is equivalent to something like "management in a given direction". If we take the two Japanese words separately, "Hoshin" means "compass needle", which in this context represents "direction", while "Kanri" means "management". This issue of direction, or vision, has been addressed several times throughout this book, as continuous improvement is also closely linked to the direction in which that same improvement should be oriented. The Hoshin Kanri concept is perfectly aligned with continuous improvement and can be described as a technique that seeks to align all the departments and sections of the organization in the same direction, the latter being defined by the top management. This means, as we say in everyday language, "getting everyone rowing in the same direction". One important part is to define the direction (*also known as the "vision"*) and the other important part is to propagate that vision throughout the structure of the organization. This second part is to make each department/section create its direction in a way that is aligned with the direction set by top management. In addition to these two parts, the Hoshin Kanri method also includes the monitoring and control mechanisms in order to ensure that the direction is being followed and that the short, medium and long-term objectives are being achieved. This subject is one of the exceptions presented in this chapter since it is presented in more detail in chapter 5.

Toyota Kata

The methodology presented by Mike Rother in his book "*Toyota Kata*" (Rother, 2010) is based on the author's interpretation of how continuous improvement is materialized at Toyota. The author puts a lot of emphasis on the invisible aspects that are imprinted in the culture and behavior pattern at Toyota. This invisible part to which chapter 5 of this book is dedicated is well represented in many principles expressed in both the Toyota Way and the Shingo Model, which focus on the social part of the socio-technical nature of business. The methodology follows the following phases:

- (1) Understand the direction: one should start by understanding where one is supposed to go, in line with the organization's vision;
- (2) Understanding the current situation: getting to know the way the process works, its standards and performance;
- (3) Define the next target condition: define well what the short-term (*a few weeks*) performance objectives are;
- (4) PDCA towards the target condition: do daily PDCA cycles with the aim of achieving small daily increments towards the target. When the target is reached, return to point (1).

This subject is another exception presented in this chapter since it is presented in more detail in chapter 7.

DMAIC Methodology

The DMAIC cycles (*Design, Measure, Analyze, Improve and Control*) (Carroll, 2013) are closely linked to the Six Sigma movement, as a mechanism for structured problem solving or improvement, a bit like PDCA cycles. This method is a little more elaborate than PDCA cycles and contains some characteristics of A3 reports. The main advantage of DMAIC cycles over PDCA cycles is in the way they guide the analyst in the planning phase, including the detailed definition of the problem, its assessment and decomposition into smaller parts. This same approach is also followed by A3 reports. Table Annex.1 presents a possible interpretation of the parallelism between the three methodologies to solve problems or achieve improvements.

Table Annex.1 Comparison between PDCA, DMAIC and A3 Report.

PDCA	DMAIC	A3 Report
(P) Plan	(D) Define	Clarify the problem
	(M) Measure	Detail the current situation
		Establish objectives
	(A) Analyze	Analyze causes
		Develop countermeasures
(D) Do	(I) Improve	Implement countermeasures
(C) Check	(C) Control	Monitor
(A) Act		Standardize

A difference that perhaps makes sense to take into account is the cyclical nature that is very pronounced in PDCA cycles, but less pronounced in the DMAIC methodology and even less so in A3 reports.

Ringi Technique

This technique is used in Toyota and other Japanese organizations to achieve decisions with broad consensus. For decisions that are reached with the consensus of all, implementation and enforcement will be easier to achieve. There is not much documentation published on this technique but there is an interesting scientific paper entitled "*"Ringi System" The Decision Making Process in Japanese Management Systems: An Overview*" (Sagi, 2015) for anyone interested in the topic.

The Ringi decision-making process is closely aligned with the Toyota Way principle "*Make decisions slowly by consensus, thoroughly considering all options; implement decisions quickly*". It is important to note that decision-making by consensus has the disadvantage of being very time-consuming when compared to majority decision-making. This apparent disadvantage makes it unpopular in the west, but if we look at the trade-off between the initial investment in time for decision-making and the effectiveness of the decision once made, we can conclude that it will be worth it in many cases. When a majority takes a decision, there will be a minority who do not agree with the decision and may resist to its implementation and compliance. That is the price to pay for that model of decision-making. In the case of the Ringi technique, the price to pay is in the time that needs to be spent until the decision is made. In addition to the advantage of improving the effectiveness of the implementation of decisions taken, there is another important aspect that gives the Ringi technique an advantage over majority decision-making. This is the fact that during the decision-making process some actors may become aware of aspects that they did not know. In this technique, a first anonymous proposal is printed out and circulated among all the stakeholders so that everyone can agree or add some detail or comment. After going through all the stakeholders, the proposal is re-drafted based on the recommendations and comments added and again circulated to everyone. This process continues until everyone agrees. This is also covered in chapter 3.

5S Technique

The 5S technique (Hirano, 1995) is probably the most popular of all those usually associated with Toyota-inspired models of excellence. It is intended to make workspaces more effective, more efficient and safer. In general terms, the proper application of this technique ensures that there is a place for everything (e.g. tools and materials) and that everything, when not in use, is always in its place. It is called the 5S technique because it involves five phases whose names, in the original Japanese words, begin with the sound "S":

- *Seiri (Sorting)* - Removing from the workspace of all items that are not needed.
- Seiton (Organizing) - Putting in appropriate places the necessary items so that they can be found easily.
- *Seiso (Clean)* - Cleaning and inspecting the workspace (*e.g. detecting an oil leak in a machine*).
- *Seiketsu (Standardize)* - Creating standards so that the workspace is kept clean and tidy.
- *Shitsuke (Discipline)* - Create routines to keep everything organized and clean.

By implementing 5S, it is possible to improve productivity, safety and quality of the products produced, as well as the work environment. The waste associated with the search for things ceases to exist and this technique allows the creation of a calmer environment, more comfortable for people and, consequently, allows an increased focus on what is important.

There is a lot of material available about 5S, namely books, videos on YouTube and texts on websites, but there are some suggestions that we can leave here:

- The initial enthusiasm will lose momentum over time.
- Do not proceed to the implementation of 5S without planning well how to solve the question of the sustainability of the technique (*the 5th S*).

Define well the mechanisms you will use to ensure that 5S routines last over time.

- There are several materials that refer to the 4th S as hygiene or health and well-being. There must be a reason why this idea has been created and copied in many places, but that is not the purpose of the 4th S. The purpose is to create standards about cleanliness, about how everything is tidied up, about 5S documentation and other related standards to make it easier for everyone.
- The purpose of 5S is to create more effective and efficient work zones in order to achieve effective performance improvements, not for the purpose of everything looking prettier and tidier.

Andon

The Japanese word *Andon* means "*light bulb*" and has come to be used to refer to a light signal indicating that a problem has occurred. In some cases, sound signals are also used for the same purpose. This concept is also linked to the ability that was given to Toyota assembly line workers to stop the line and automatically trigger an *Andon* light signal to let everyone know quickly that something is wrong somewhere on the line. *Andon* systems have evolved over time and today you can find complex systems with monitors showing, for example, the deviation of production from the plan.

A very common type of *Andon* system is a set of different colored lights as shown in the workstation represented in Figure Annex.3. Typically, these sets of lights show the status of the equipment with a coding such as: green - running normally, red - production stopped, blue - out of material, yellow - needs servicing. These simple systems, like many others, prove to be extremely effective.

Kamishibai

The Japanese word *Kamishibai* refers to a type of street theatre in which typically a person tells a story with the help of a support of images related to the story. Each colored drawing that is shown to the audience has on its back the text, or notes, that the actor uses to help him tell the story without forgetting anything (*equivalent to the "prompter" in theatre*). As far as the production context is concerned, the word *Kamishibai* was adopted to designate a simple visual technique that is typically used to inform whether a given activity or operation that needs to be carried out in the period in question has already

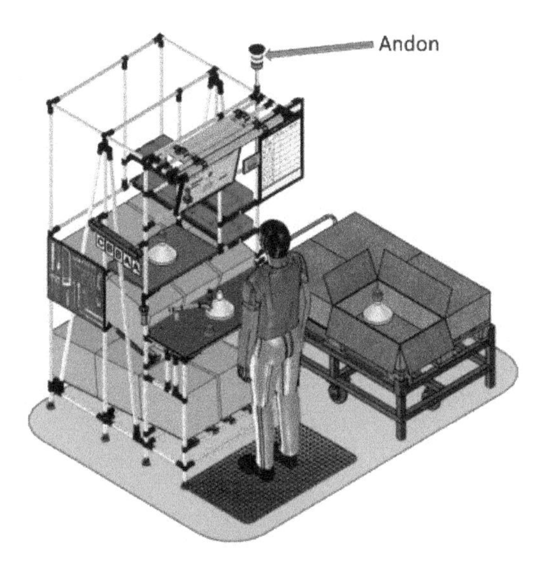

Figure Annex.3 Example of an Andon device (courtesy of 4Lean.net).

been done or is still missing. This technique is based on a board with cards that are green on one side and red on the other. If the exposed side of the card is green it means that the operation in question has already been carried out and, if it is red, it means that the operation has not yet been carried out. The example shown on Figure Annex.4(a) shows the use of this technique as a way of ensuring that maintenance tasks are effectively carried out without any being forgotten. Let us see how it works; at the end of each day all cards corresponding to daily tasks must show the green side, indicating that all tasks have been carried out. Then, all the cards are turned to the red side (*a sort of resetting of the system*) so that the next day the day begins with all the tasks marked as "to do". If at the end of a day, any of the tasks still "shows" as red, then someone forgot to do it and it is necessary execute it, before doing the aforementioned "reset".

There are other uses for this technique, as shown in the example on Figure Annex.4(b). In this case (*courtesy of Lipor),* the *Kamishibai* card is used to control audits of workspaces.

Poka-Yoke

Poka-Yoke devices serve to prevent errors from occurring, or in other words, they are error-proof devices. This concept is very important because in many cases, with a small device, we could prevent errors that lead to the creation

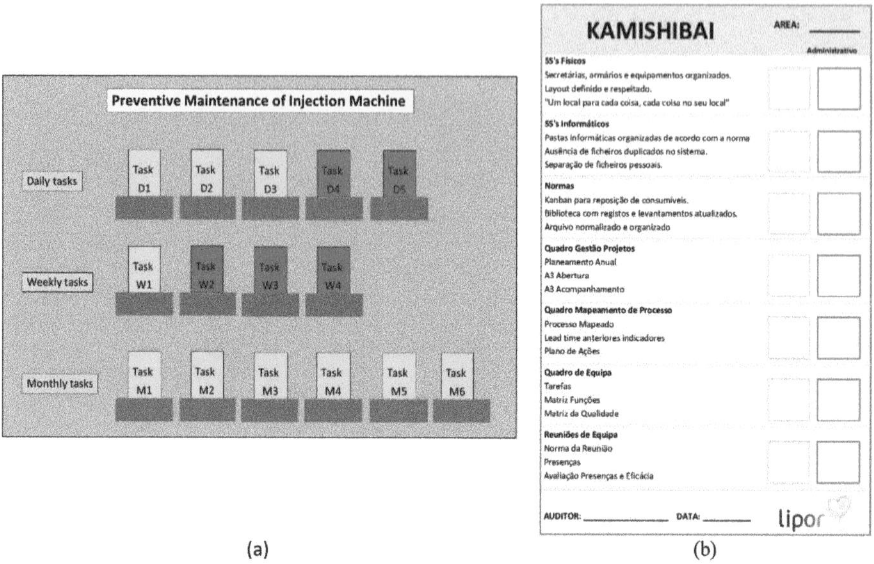

(a) (b)

Figure Annex.4 Example of Kamishibai in (a) maintenance (b) audits at Lipor (courtesy of Lipor).

of defects, incidents or accidents, many of them with fatal outcomes. Poka-Yoke systems are very useful in industry, but can also be used effectively in our daily lives. If the reader is interested in deepening their knowledge of this concept here are two old but enlightening books: the book "*Poka-Yoke: Improving Product Quality by Preventing Defects*" from 1989 (Shimbun, 1989) and the book "*Mistake-Proofing for Operators: The ZQC System*" from 1997 (Productivity Press Development Team, 1997).

The examples of Poka-Yoke solutions presented in Figure Annex.5 are the following:

a. The pin that becomes part of the support does not allow the piece to be placed in the wrong way.

b. In the past, when one went to fill up at the fuel pumps it was common to put the cap of the tank on top of the car roof while the car was being filled up. Sometimes the driver would forget to put the cap back on and the result was quite unpleasant. With the small cable that holds the cap as shown in photo a) of Figure Annex.5, this mistake was largely avoided.

Figure Annex.5 Examples of Poka-Yoke devices.

c. This case is an example of the kind of approach we can take in our daily lives. If a person has to put on that shoe to leave the house, they will not forget their phone.

These are just a few examples, but the possibilities are endless for the use of these devices and solutions in industry, especially with the introduction of electronics and information systems.

Jidoka

The concept of *Autonomation* was also presented by Taiichi Ohno (Ohno, 1988) as one of the two pillars of the Toyota Production System. This concept, originally called *Jidoka* (*Japanese word*), was in fact introduced by Sakichi Toyoda (1867-1930), a great Japanese inventor and entrepreneur, initially dedicated to loom manufacturing and later linked to the foundation of Toyota. One way of describing it, proposed by Yasuhiro Monden (Monden, 1998), is that *Jidoka* prevents defective units of a preceding process from flowing and disrupting the next process. The idea is to introduce sensors and logic into machines to ensure that the machine automatically stops if any error occurs. This machine capability (*Jidoka*) eliminates the need for a person to monitor the equipment and intervene only if something goes wrong. People spend their time on tasks with greater added value, where they can use more advanced skills than simply monitoring equipment. Poka-Yoke systems, when applied to stop a piece of equipment or warn the employee when something is wrong with a piece of equipment, are serving the *Jidoka* purpose.

Overall Equipment Effectiveness (OEE)

This indicator, initially described as a central component in the TPM (*Total Productive Maintenance*) methodology (Nakajima, 1988) assumes that in a certain, apparently ideal, point of view, the equipment should produce at 100% of its potential. Let us see the following example: if a piece of equipment has the capacity to produce 100 pieces per hour, then in an 8-hour shift it should ideally produce 800 pieces in good conditions. From the point of view of equipment utilisation and financial profitability, this would be the ideal situation, but this is never the reality. There are, however, a number of factors that prevent this from being possible and OEE is an indicator that addresses these realities in a very effective way. Overall, OEE indicates what percentage of a piece of equipment's time was actually used to produce products in good condition as the following equation shows.

$$OEE = \frac{Quantity\,of\ good\ products\ produced}{Theoretically\ possible\ production\,quantity} \times 100\% \qquad \text{(eq.1)}$$

The equation (eq.1) gives us the relationship between the quantity of fault-free products that the equipment produced in a given period of time and the quantity that, in theory, it would be possible to produce in that same period of time. However, the OEE goes a little further by identifying what kind of factors weigh on this value. Firstly, some of the time that the equipment should be producing is used for necessary rest breaks for the operator(s), planned maintenance stoppages and/or simply for lack of demand. Secondly, unplanned events such as breakdowns, setups, raw material jams, material shortages, etc. occur throughout a shift. Moreover, during the time that the equipment is actually producing, we still have the cases where the equipment does not work at its ideal pace because it is already worn out, the tools are worn out, the operator is inexperienced, the raw material is not the best, etc. Finally, we still have the problem that is the production of defective product. Therefore, the OEE is actually broken down into three partial indicators as shown in equation (eq.2):

$$OEE = Availability * Performance * Quality \qquad \text{(eq.2)}$$

where:

$$\text{Availability} = \frac{\textit{Shift time} - \textit{planned stops} - \textit{unplanned stops}}{\textit{Shift time} - \textit{planned stopps}} \qquad \text{(eq.3)}$$

$$\text{Performance} = \frac{\textit{Ideal cycle time} * \textit{Quantity of products produced}}{\textit{Shift time} - \textit{planned stops} - \textit{unplanned stops}} \qquad \text{(eq.4)}$$

$$\text{Quality} = \frac{\textit{Quantity of good products produced}}{\textit{Quantity of products produced}} \qquad \text{(eq.5)}$$

Apart from the details of calculation and the different approaches that are adopted in different organizations, some aspects should be taken into account. One of them is that monitoring the OEE is important in some equipment but not in all. In fact, it is in the equipment that limits the organization's production, the bottlenecks according to Eliyahu Goldratt's Theory of Constraints (Goldratt & Cox, 1984), that one should monitor and continuously try to improve the respective OEE. It would be important to know the flow of value within the organization in order to better judge which equipment to monitor in terms of OEE. Another aspect to take into account is how the OEE should be monitored. Constant automatic data collection is the best way to do it.

Value Stream Mapping (VSM)

This is one of the most popular techniques or tools used in "Lean" initiatives carried out in organizations. The tool is used to graphically represent the flow of materials through the various processes in an organization, from the purchase of materials and raw materials to dispatch to customers. It is a representation tool that includes the main information flows between the production planning and control system and not only the organization's main processes, but also the main suppliers and customers. The most famous books on the subject are: "*Learning to See: Value Stream Mapping to Add Value and Eliminate MUDA*" by Mike Rother and John Shook (Rother & Shook, 1999) and "*Seeing the Whole: Mapping the Extended Value Stream*" by Daniel Jones and James Womack (Jones & Womack, 2002). An example of a VSM is presented in Figure Annex.6.

It is important that several people can be discussing around a VSM that represents their reality in order to analyze which are the main obstacles in the

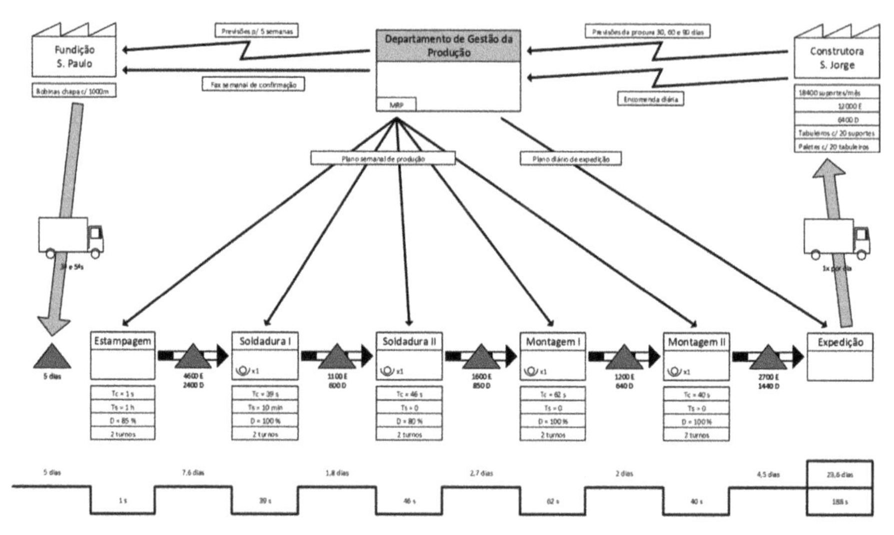

Figure Annex.6 Example of a VSM. Adapted from Rother and Shook (1999).

flow that may be preventing to do better and try to establish an action plan for a better future state. One of the most relevant aspects is the identification of flow gaps, or in other words, the stagnation times of materials between processes. These times are estimated using Little's law presented in chapter 6, and there are often doubts about how to account for them. Intuition leads us to think that the time that items wait in a process should be a function of the cycle time of that process, but, in reality, this is not the case. In fact, the waiting times in each process should be calculated by multiplying the corresponding WIP with the *Takt* time of the market.

Process Mapping

Process mapping is a designation that is assumed here to be the process of representing information flows in the indirect areas of organizations. Let us say that process mapping would be the equivalent of VSM, but aiming to represent information flows only. The application of VSM is not feasible in the indirect areas of organizations, given the complexity of information circuits between the different stakeholders. Let us take the example presented in Figure Annex.7, which shows a process mapping carried out in a cable manufacturing organization regarding the consultation process for new projects.

As you can see, a long wall was required so that the whole process could be represented in a single diagram. A customer consultation process

Figure Annex.7 Example of process mapping.

goes through several departments in the organization and with possible repetitive exchanges of information between departments and sometimes consulting the customer during the process. The process of bringing together in the same room, representatives from all the departments involved to build such a map, is very important because it helps to understand better the effect that the actions and behaviors of each stakeholder have on overall performance. This awareness is extremely important in order to discover redundancies, errors, rework, etc., that only by looking at the process as a whole can be seen. Many are the improvement proposals that can be identified and many gains can be achieved for the organization and for the intervenient themselves.

Spaghetti Diagrams

Spaghetti charts have that name because what results in the charts resembles spaghetti, as shown in the green line on Figure Annex.8. The idea is to draw, on the floor plan of a production unit, which you want to analyze, a line that represents the movements of one or more employees during a certain period. It would be ideal if the employee carried a device that constantly uploaded his position to the system. With that record, a line could be drawn on the floor plan as if it were the employee's trail. As this type of technology is not always easily accessible, one way to do this is to take a sheet of paper with the layout of the site and mark on it the path similar to the movements of the employee. This work is not easy because it requires the analyst to observe constantly the movements of the employee during a certain period (*for example one*

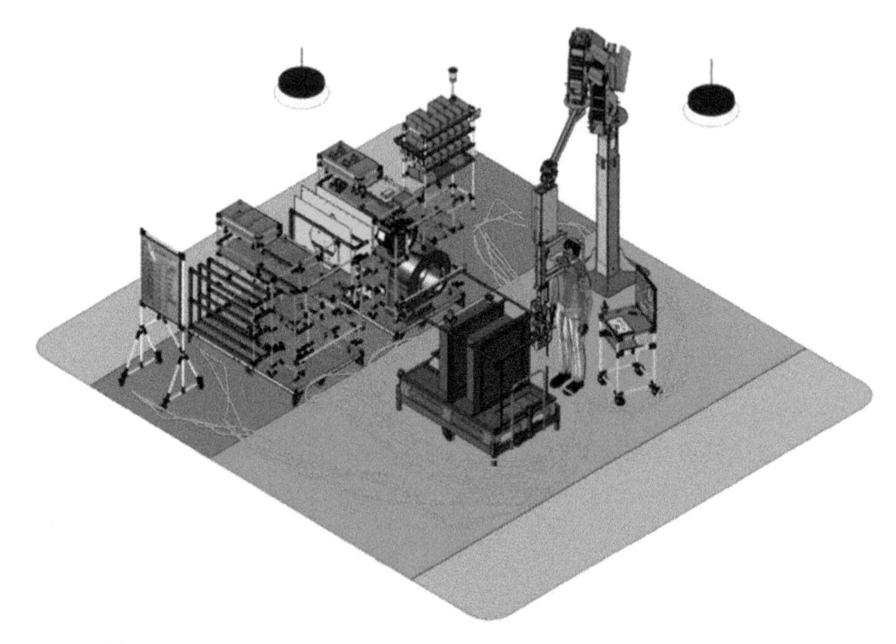

Figure Annex.8 Example of Spaghetti diagram (courtesy of 4Lean.net).

shift). The aim with the result of the analysis is the clear demonstration of the incidence of the movements. In view of the areas or paths with the highest incidence, the analyst may look for different arrangements of equipment or changes in working procedures and standards.

SMED

Single Minute Exchange of Die (SMED) is a technique that Shigeo Shingo started to develop at Mazda in 1950 and presented in 1985 in his book "*A Revolution in Manufacturing: The SMED System*" (Shingo, 1985). The name of this method is not very intuitive nor is it very easy to understand its meaning. The term "*Single Minute*" is intended to translate the idea of "*single digit in minutes*", that is, a time up to a maximum of 9 minutes to be achieved using this methodology. One of the first lessons one learns, as one works on projects to reduce equipment setup times, is that a large part of this time can be reduced just by changing the way the setup operations are performed. Often, no change in the equipment is required to achieve major savings in setup time. This technique starts by recording the whole setup process, ideally using video recording, and then analyzing the detail of all the operations that are carried out. Each of these operations is then classified as being an

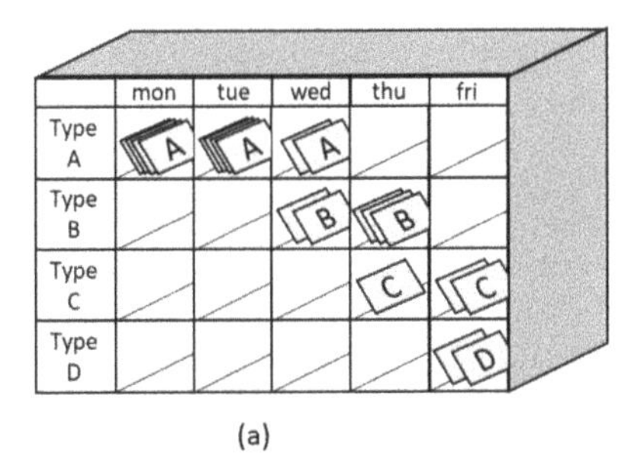

(a)

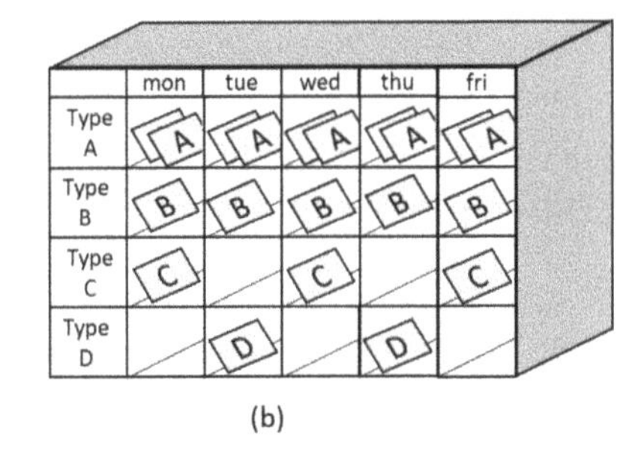

(b)

Figure Annex.9 Heijunka box.

internal operation or an external operation. Internal operations are those that can only be performed while the machine is stationary and external operations are those that can be performed while the machine is running. Fetching tools, taking tools, approaching raw materials, preparing tools are examples of external operations. Putting the tool in the machine or loosening supply guides are examples of internal operations. All operations classified as external must be then carried out either before stopping the equipment or after restarting the equipment *(after carrying out the setup)*. Then, there is the possibility to include some devices for faster tightening and tuning and to include more people in the equipment preparation operations.

Heijunka

Heijunka's concept (*levelling*) is to achieve, as best as possible, stability and consistency in production. Heijunka is aligned with two important concepts or principles practiced at Toyota, which are fluidity in production and the elimination or reduction of Mura (*variability*), one of the 3 enemies of production, presented in chapter 2. The evolution from the type of approach to production scheduling presented in Figure Annex.9(a) to the type of scheduling approach in Figure Annex.9(b) is a materialization of *Heijunka's* concept.

In order to clarify the idea it will be better to clarify the following: each card that appears in Figure Annex.9, represents a batch of articles of a specific quantity (*in the limit it can be a batch of only one unit*). In Figure Annex.9(a), it is programmed to produce on Monday and Tuesday 4 batches of article A. On Wednesday, it is finished the production of all the articles A necessary (*two batches*) and it is still programmed to produce two batches of article B and so on. This traditional approach is based on the understanding that you first produce all As and then all Bs and so on. Changing products is not something that managers naturally like to do.

It is important to note that although on both sides of the Figure Annex.9 a *Heijunka* box is used as a way to present the production programs in both approaches, this type of box is only used when the purpose of the *Heijunka* concept is intended.

Let us then look at some differences between case a) and case b) in Figure Annex.9. In the first case, since not all products are produced every day, it is necessary to keep all types of products in stock to meet the eventual demand on any day of the week. In case b) every day we have products A and B to meet the demand, it is only necessary to stock products C for one day and it is only necessary to stock products D for a maximum of two days.

The next step, in order to follow the *Heijunka* concept, would be to start producing all types of products every day according to the average daily demand. For product A and B there was no problem but for the other products, or product types, we would have to reduce the lot size to meet one day's demand only. As the *Heijunka* concept is refined in an organization the more stable and constant is the supply of materials, both from internal and external suppliers. The constancy and stability of ordering from suppliers for materials supply makes everything simpler for them.

The book with the title "*Creating Level Pull: A Lean Production-System Improvement Guide for Production-Control, Operations, and Engineering Professionals*" (Smalley, 2004) is an example where the reader can delve into this and other tools for pull production.

Kanban Systems

The *kanban* systems, which were developed within the framework of TPS, materialize the famous *Just-In-Time* concept, which aroused great curiosity among western managers and academics in the 1970s and 1980s. The first work to devote much attention to it was probably the Japan Management Association's book "*Kanban Just-in Time at Toyota: Management Begins at the Workplace*", published in 1985 and given its English language version by Productivity Press (Japan Management Association, 1986). The subject is also discussed extensively in Shigeo Shingo's famous book on TPS (Shingo, 1989). *Kanban* systems are a visual and simple way of controlling the flow of materials for production or transportation that are very economical in an industrial environment. In its most classic form, *kanban* cards function like the "*token*" that athletes in relay races use. In those races an athlete waits until the "*token*" has been handed to them, and only then is he/she allowed to start his/her part of the race. In an industrial environment, a *kanban* card takes on this token role. A box or a trolley with parts can only move on to the

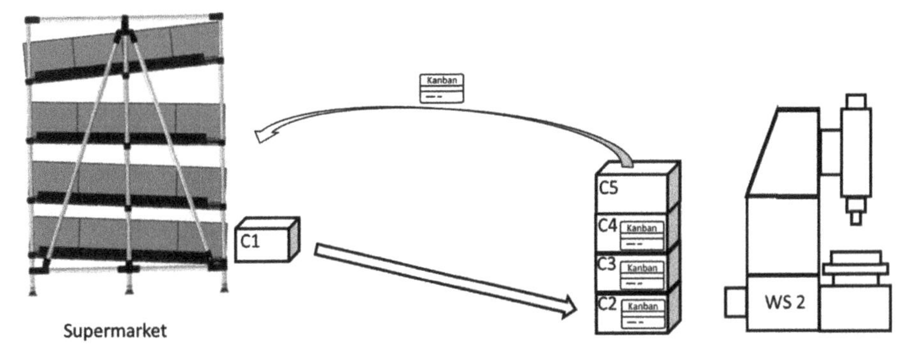

Figure Annex.10 Kanban system. Image of supermarket (courtesy of 4Lean.net).

next process if an authorization is given to it in the form of a *kanban* card as shown in Figure Annex.10.

As an example, the transport *kanban* circuit shown in Figure Annex.10 works as follows: in this case there are 4 *kanban* cards circulating between the supermarket and workstation WS 2; container C1 cannot move to WS 2 because it does not yet have authorization, which is given by the existence of a *kanban* card; when the parts from container/box C5 have started to be processed at WS 2, the *kanban* card is available to be sent to the supermarket; when this card arrives at the supermarket, container C1 can be moved to WS 2. In this case, it is a transport *kanban* but the same reasoning is used for production. The calculation of the number of *kanbans* that must circulate between the supermarket and WS 2 for the parts in question depends on the consumption of these parts and the time it takes for a container to arrive after a *kanban* is released at WS 2. This calculation is developed in chapter 5.

There are other ways to implement production and transport authorizations following the *kanban* logic such as marked spaces on the floor or on benches. If the space is empty, it is giving permission for a new part or a new box with parts to fill that void. Digital systems can also be used for the same purpose.

Lineside Rack

The lineside rack is the interface between logistics and production (*although it is called lineside rack, this concept is also used in workplaces that are not part of a line*). Lineside rack is a designation that may not be used in many organizations but is proposed in the book "*Total Flow Management*" (Coimbra, 2009) and that, in our opinion, suggests, with some effectiveness, its function. It is the place where the parts and components necessary for

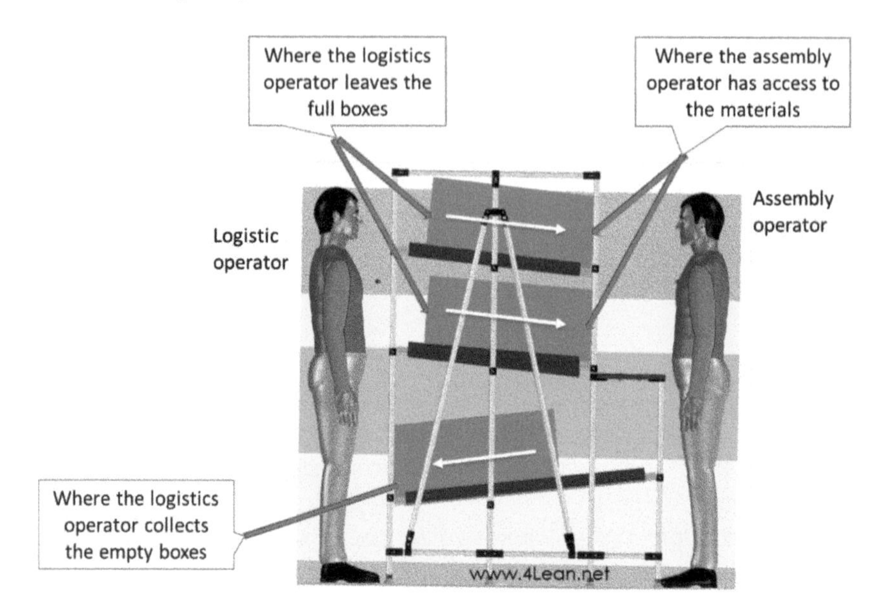

Figure Annex.11 Schematic of a line side (courtesy of 4Lean.net).

assembly (*or manufacturing*) are deposited by logistics in order to be within reach of the operator at the manufacturing or assembly stations. The lineside rack must be designed so that both the production operator and the logistics operator can, respectively, collect and deposit the materials/components with the least possible effort (Figure Annex.11). Parts and components can, and are in some cases, placed on pallets on the floor near the production or assembly operator but this is not a good solution because it creates a lot of waste with operator movements. To improve productivity, one of the most popular line edge typologies is the use of dynamic racking (inclined roller racks, which operate by gravitational action) as shown in Figure Annex.11.

The production or assembly operator can effortlessly gain access to the parts or components that are typically in containers/boxes positioned in the live racking on the operation side. On the other side of the dynamic rack, the logistics operator fills the full boxes and removes the empty boxes. This way, there is no interference/collision between the production operator and the logistics operator, so this is one of the best types of line edge.

Logistics Train

Logistic train is one of the possible names, although it can also be known as *Mizusumashi* or *Milk Run*. The Japanese name *Mizusumashi* comes from

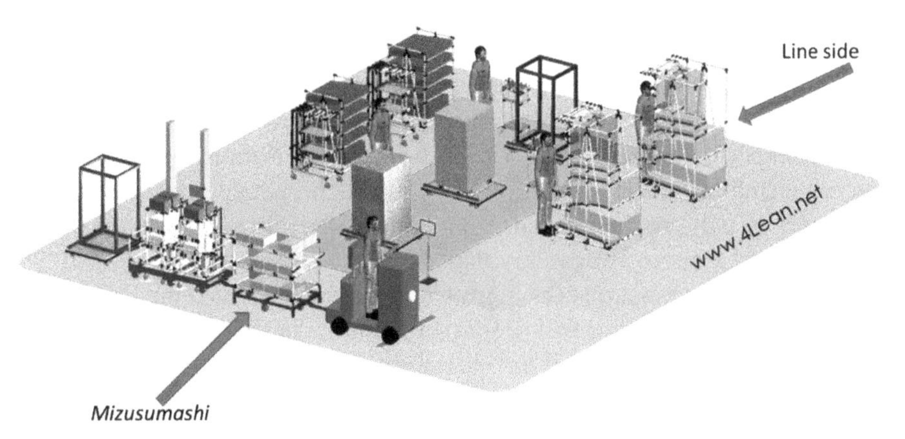

Figure Annex.12 Mizusumashi in the production context (courtesy of 4Lean.net).

the insect "water spider" that walks on water in lakes and rivers. One of the explanations for its use in this context is related to the fact that the water spider walks between water and air, in analogy with the fact that the logistics train moves between production and logistics. It does not sound like a very good analogy, but it is one of the known explanations. As for the term *Milk Run*, its analogy is more logical. In some countries, such as England, the milkman runs along the same route every morning, replacing the empty milk bottles left at the door of each house on that route with an equal number of full bottles of milk. It is a very simple and clear system that works very well. If a person wants three bottles of fresh milk in the morning, they only have to leave three empty bottles the night before outside the door. The operator of the logistic train, whose appearance is usually similar to the representation shown in Figure Annex.12, acts according to the same logic, replacing empty boxes he finds on the lineside racks by full boxes of components. The major paradigm difference introduced with the logistics train in internal logistics is the following: The traditional form of logistics within factories is to have transport between different origins and different destinations as it is needed. It is even common for the production operator, when faced with the need for components, to go himself or ask the logistics staff to supply him, with production losses while the supply is carried out.

In this new approach, the logistics train, or *mizusumashi*, functions like an underground train, passing at scheduled times in all stations. In each cycle, the logistics operator only has to replace the empty boxes he found in the previous cycle with full boxes, on all the lineside racks (Figure Annex.12) that are part of his route. Very often, the cycles of these logistic trains are of 20 minutes or one hour.

Supermarkets

The term "*supermarket*" is used to designate a specific type of storage because of its similarities to the supermarkets we are used to shopping in. Supermarkets, in the context of material flow management in industrial units, are organized intermediate storage spaces, which operate according to the principles of pull production. This concept is different from traditional storage in several aspects. We can say that supermarkets stand out in the following aspects: they ensure that the items which have been stored the longest are those which are consumed first (*FIFO discipline*); they allow visual management (*it is easy to know what the maximum stock and the order point are*); they are easy to access for picking (*an operator has easy and equipment-free access to all stored items);* items and components are stored in small boxes which, in some cases, may have wheels (*if they are on the floor*); and there is a single location for each item/component.

The FIFO discipline is achieved because in most cases boxes are deposited on one side of the dynamic chutes and collected on the other side, as shown in Figure Annex.13. The full boxes get in the supermarket on the left side and are picked up by the logistics operator on the right side to be transported to the lineside racks. The boxes slide over chutes with rollers by gravity. Empty boxes returned by the logistics operator are placed on racks with designated chutes. In Figure Annex.13, this happens at the top of the rack but there are also cases where the rails are used at the bottom of the rack.

The logistics operator passes with the logistic train through the supermarket, in each of its cycles, on the picking side, as shown in Figure Annex.14.

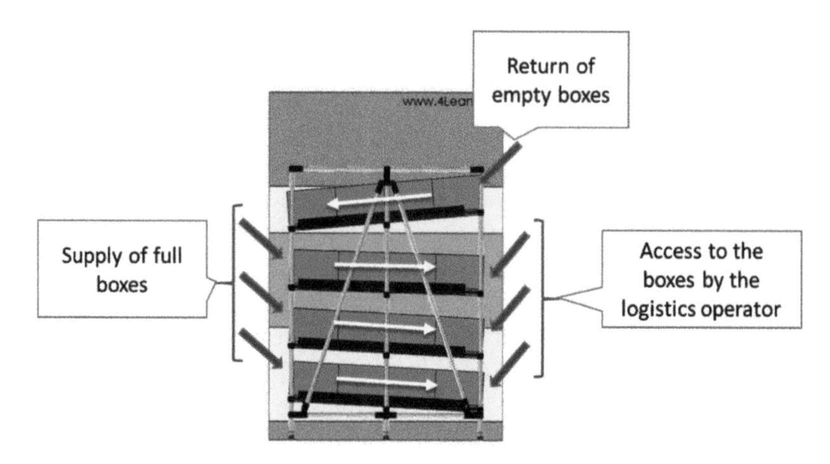

Figure Annex.13 Scheme of operation of a supermarket (courtesy of 4Lean.net).

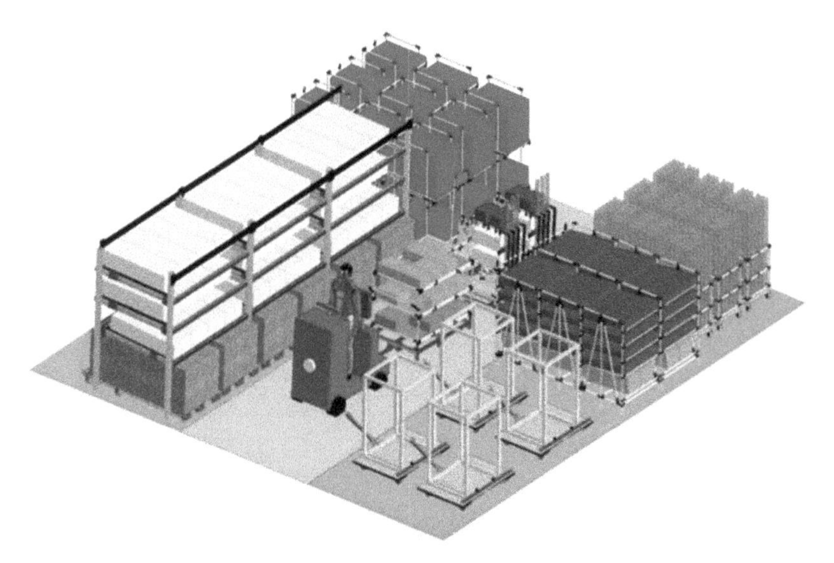

Figure Annex.14 Supermarkets in industrial context (courtesy of 4Lean.net).

When passing through the supermarket, the logistics operator returns the empty boxes, using the chutes defined for that purpose, and supplies himself with full boxes, which he needs to take to replace the empty ones at the lineside rack. The authorization to replenish materials in the supermarket is usually triggered by the order point defined for each case, as described in chapter 5.

References

Bicheno, J. (2015). *The Lean Games and Simulations Book*. Picsie Books.

Carroll, C. (2013). *Six Sigma for Powerful Improvement: A Green Belt DMAIC Training System with Software Tools and a 25-Lesson Course*. CRC Press.

Coimbra, E. A. (2009). *Total flow management : achieving excellence with kaizen and lean supply chains*. [Howick, N.Z.]: Kaizen Institute.

Goldratt, E., & Cox, J. (1984). *The Goal*. Great Barrington: North River Press.

Hirano, H. (1995). *5 pillars of the visual workplace: The source book for 5S implementation*. Productivity Press.

Ishikawa, K. (1985). *What is Total Quality Control? The Japanese Way*. Englewood Cliffs: Prentice-Hall.

Japan Management Association. (1986). *Kanban Just-in Time at Toyota: Management Begins at the Workplace*. Productivity Press.

Jones, D., & Womack, J. (2002). *Seeing the Whole: Mapping the Extended Value Stream. Lean Enterprise Institute, Brookline*. Cambridge, MA, USA: Lean Enterprises Inst Inc.

Matthews, D. (2010). *The A3 Workbook: Unlock Your Problem-Solving Mind*. CRC Press.

Monden, Y. (1998). Total Framework of the Toyota Production System. In *Toyota Production System: An Integrated Approach to Just-In-Time*.

Nakajima, S. (1988). *Introduction to TPM: Total Productive Maintenance*. Productivity Press.

Ohno, T. (1988). *Toyota production system: beyond large-scale production*. (C. Press, Ed.) (3ª Edição). New York: Productivity, Inc.

Productivity Press Development Team. (1997). *Mistake-Proofing for Operators: The ZQC System*. Productivity Press.

Rother, M. (2010). *Toyota KATA: Managing People for Improvement, Adaptiveness and Superior Results*. McGraw-Hill Education - Europe.

Rother, M., & Shook, J. (1999). *Learning to see: Value stream mapping to add value and eliminate muda. The Lean Enterprise Institute*. https://doi.org/10.1109/6.490058

Rundle, R. (2019). *Deming Cycle Pdca - Plan Do Check ACT Journal in Daily Life Toyota Way*. Independently Published. Retrieved from https://books.google.pt/books?id=vxDjwgEACAAJ

Sagi, S. (2015). "Ringi System" The Decision Making Process in Japanese Management Systems: An Overview. *International Journal of Management and Humanities*.

Shimbun, N. (1989). *Poka-Yoke: Improving Product Quality by Preventing Defects*. Productivity Press.

Shingo, S. (1985). *A Revolution in Manufacturing: The SMED System*. Oregon: Productivity Press.

Shingo, S. (1989). *A study of the Toyota production system from an industrial engineering viewpoint*. New York: CRC Press.

Smalley, A. (2004). *Creating Level Pull: A Lean Production-System Improvement Guide for Production-Control, Operations, and Engineering Professionals*. Lean Enterprise Institute.

Index

About the Author

José Dinis Carvalho was born near Barcelos in the north of Portugal, having spent 4 years of his childhood in Angola. He graduated from Minho University with a degree in Production Engineering in 1989, completed an MSc at Loughborough University in Computer Integrated Manufacturing in 1992 and completed a PhD at Nottingham University in 1997. He continued as a lecturer at the Production and Systems Department of the School of Engineering in the University of Minho, teaching subjects in the areas of Production Organisation and Management. From 2004, with the help of his colleagues, he started to introduce active learning practices, in particular project-based learning practices in several engineering courses. These initiatives led him to win a merit award for teaching in 2009. He founded with his colleagues an association on project-based learning called PAEE (Project Approaches in Engineering Education) that has been organising to date, conferences on the subject in several countries around the world. It was also from 2004/2005 that he started to dedicate himself more deeply to the study and application of concepts and principles of Lean Philosophy and operational excellence in companies through projects with students and colleagues, which dictated his path as researcher and teacher. At the moment he is an associate professor in the same department and his research focus is continuous improvement and excellence in organisations.

Collaborator Rui M. Sousa was Born in Matosinhos in 1966. He graduated in Electrical Engineering (telecommunications and electronics branch) from the University of Coimbra in 1989 and concluded, in 1996 and at the same University, the Master's Degree in Systems and Automation (industrial automation branch). In 2003, he finished his PhD in Production and Systems Engineering at the University of Minho. He was a professor at the Higher School of Technology and Management at the Polytechnic Institute of Leiria between 1993 and 1996. In 1996, he joined, as assistant teacher, the Department of Production and Systems at the School of Engineering at the University of Minho, where he is currently Associate Professor.

He teaches in areas such as continuous improvement and production organization and management, although at the research level his focus is on the area of continuous improvement, with involvement in several projects in the company. He has also been working since 2004 with innovative forms of training, namely active learning involving gamification, both academically and professionally.

For Product Safety Concerns and Information please contact our EU
representative GPSR@taylorandfrancis.com
Taylor & Francis Verlag GmbH, Kaufingerstraße 24, 80331 München, Germany

www.ingramcontent.com/pod-product-compliance
Ingram Content Group UK Ltd.
Pitfield, Milton Keynes, MK11 3LW, UK
UKHW021821240425
457818UK00001B/15

* 9 7 8 8 7 7 0 2 2 7 9 8 8 *